Biography of an Industrial Landscape

Biography of an Industrial Landscape

Carlsberg's Urban Spaces Retold

Svava Riesto

Routledge
Taylor & Francis Group
LONDON AND NEW YORK

First published in 2018 by Amsterdam University Press Ltd.

Published 2025 by Routledge
4 Park Square, Milton Park, Abingdon, Oxon OX14 4RN
605 Third Avenue, New York, NY 10158

Routledge is an imprint of the Taylor & Francis Group, an informa business

ISBN: 9789089647351 (hbk)
ISBN: 9781041176183 (pbk)
ISBN: 9781003691624 (ebk)

NUR 682

Cover illustration: Visualization from the winning proposal of the ideas competition 'Carlsberg Vores by' 2007. Entasis Architects/Carlsberg Ltd Properties.

Cover design: Coördesign, Leiden

DOI: 10.5117/9789089647351

Made with the support of

CENTER FOR STRATEGISK BYFORSKNING

TUBORG FONDET

For Product Safety Concerns and Information please contact our EU representative:
GPSR@taylorandfrancis.com
Taylor & Francis Verlag GmbH, Kaufingerstraße 24, 80331 München, Germany

Table of Contents

Acknowledgements

This book emerges from inspirational exchange with many knowledgeable and generous people, first and foremost Malene Hauxner, Ellen Braae and Tom Avermaete, who supervised my Phd project on the Carlsberg brewery redevelopment from 2008 to 2011. Colleagues at the Landscape and Planning Department of the University of Copenhagen as well as the Öresund Design Research Seminars provided a stimulating ground for testing ideas and giving birth to new ones; thank you, Lisa Diedrich, Anne Tietjen, Henriette Steiner, Maria Hällstrøm Reimer and Vera Vicenzotti. Jan Kolen, Andrea Kahn and Søren Kjørup provided insightful comments at crucial moments in the research.

I want to thank the people at the Culture Agency of Denmark (formerly the Heritage Agency), in particular Caspar Jørgensen; Carlsberg Ltd and later the Development Company Carlsberg Byen for their genuine openness towards me as a researcher. The many workers that I encountered during the studies on-site have been extraordinary guides to knowledge as they shared their experiences during walks and talks in the breweries. Also, the Copenhagen Municipality, Copenhagen Museum, Frederiksberg Stadsarkiv, The Danish National Archives and the Danish Worker's museum shared documents and insights. Many designers and others involved in the Carlsberg redevelopment project have been informative sources; in particular Signe Cold from Entasis; Rainer Voss, Vogt Landscape Architects, Stig L. Andersson, SLA, and Keingart. A special thanks to Ulla Nymand at Carlsberg Ltd Archives for supplying the project with records and insights that have made this work a lot of fun.

I wish to thank the University of Copenhagen, the Heritage Agency of Denmark and Carlsberg Ltd for funding of the Phd-fellowship, which enabled me to study Carlsberg's early redevelopment closely. The Centre for Strategic Urban Resarch and the Tuborg Foundation generously supported the making of this book. I wish to thank the Amsterdam University Press and the Landscape and Heritage Series Group for this opportunity to collaborate, as well as anonymous peer reviewers for very helpful comments.

Not the least I want to thank my family for all their love and support.

Introduction

'Life must be lived amidst that which was made before.'
– Donald W. Meinig[1]

'[H]eritage in fact has very little to do with the past but actually involves practices which are fundamentally concerned with assembling and designing the future.'
– Rodney Harrison[2]

When we attempt to shape the city of tomorrow, starting from a blank sheet of paper is doomed to failure. Urban planning and design are not, as some modernists may have dreamt, invention in an empty space but rather imply intervening in existing urban landscapes. Such a premise raises fundamental questions about value: What aspects of the urban landscape do we deliberately want to retain or strengthen and what to change and leave behind in striving to meet what we believe will be tomorrow's needs and desires? What kind of a reading of the existing landscape, inherited from the past, underlies such choices?

One of the key tasks of contemporary urbanism is to address the urban landscapes that were shaped for industrial purposes.[3] In the last decades former industrial production and distribution facilities in many European and North American regions have closed due to changing global and local economic, political and social processes. Finding out what roles former plants and distribution landscapes should have in the city is a new challenge in many metropolitan areas and this task is entangeled with questions about if and how to transmit the inheritance of industrial cultures of the past. Post-industrial landscapes mutate without authorized uses and official plans; materials decay, people enter without official permission, trees penetrate built structures and different human sub-cultures and faunas evolve. Those landscapes later or never become sites for explicit plans, turning them into museums, cultural centres or loft appartments, or – as is the focus of this book – comprehensive urban redevelopment plans may seek to transform

1 Meinig, *The Interpretation of Ordinary Landscapes*, 44.

2 Harrison, 'Beyond "Natural" and "Cultural" Heritage', 35.

3 See e.g. Kirkwood, ed. *Remanufactured Sites: Rethinking the Post-Industrial Landscape* and Braae, *Beauty Redeemed.*

larger industrial plants into urban districts, seeking to combine multiple urban functions such as housing, retail or cultural institutions. Such urban redevelopment often happens in regions characterized by economic and demographic growth or with aspirations to grow. Urban redevelopment projects are complex inter-disciplinary processes, which always involve processes of re-reading and re-imagining the industrial landscape while seeking to convert it to something new.

Among the key professional voices in urban redevelopment projects are architects, landscape architects, urban designers, urban planners and experts in cultural heritage. People from those fields are central actors in surveying the present situation and proposing what should be changed or preserved, reused and altered in order to best accommodate what is identified as desirable futures. Although designers, planners and heritage experts sometimes present their surveys and proposed actions as more or less self-evident – as in 'we just take care of the most indispensable values' or 'we change what obviously has to be changed' – their choices are all but innocent. Rather, the people that are professionally involved in urban redevelopment make active choices that rely on certain lenses that allow them to see and value some parts of the existing urban landscape, while leaving others in blind spots or rejecting them. While such selective lenses are a condition – no one has access to the totality of a landscape –they are not always laid open, which often hinders an open discussion and narrows the spectrum of possible ways of imagining both the pasts and futures of a post-industrial redevelopment site.

This book seeks to articulate such lenses and to discuss the more or less hidden assumptions and narratives that they rely upon in the context of contemporary urban redevelopment of former industrial production sites. By unfolding urban redevelopment as a cultural practice, I want to contribute to a critical discussion about how we value and change post-industrial landscapes during their redevelopment planning. If we accept the possibility that there is more than one answer to the question of what are the aspects of landscapes shaped by industrial production and distribution worth retaining or strengthening, we need to stimulate an explicit debate about the values that underlie our actions – especially those values and thought-frames that are taken for granted in daily practice. Potentially, it is my hope, an open discussion about values and narratives can create a richer ground of possible interpretations on which we as a society can decide how we manage change in industrial landscapes.

The redevelopment of post-industrial landscapes increasingly brings people from many disciplines together. Spatial planning and heritage making

Image 0.1 What aspects of industrial urban landscapes do we deliberately want to retain or strengthen and what to change and leave behind in striving to meet what we believe will be tomorrow's needs and desires? Demolishion of former bottlery builidng at the Carlsberg Breweries in 2013.

are more and more intertwined in processes that seek to use heritage as an asset for the planning process. This means that experts from different fields that work with heritage management, spatial planning and design increasingly work in interdisciplinary processes. Yet, trained in different disciplinary traditions, which have historically had much less contact with each other than in the present, professionals sometimes use methods, lenses and knowledge that seem inaccessible to people from other fields. I want to unveil and discuss the different positions that planners, designers and heritage managers use in their evaluations of industrial landscapes during redevelopment. My idea is that such an articulation of positions can help substantiating future negotiation processes and to contribute to the knowledge-migration across disciplinary borders.

This book studies a specific location and transformation process: Carlsberg in Copenhagen, a former brewery and a significant urban redevelopment site in northern Europe at the start of the twentyfirst century. As I write,

Image 0.2 Carlsberg is located within the dense city of Copenhagen. Aerial photograph of the site prior to the urban redevelopment project, when the Carlsberg site was a mix of buildings of different sizes, groups of metal tanks, paved surfaces and gardens.

Carlsberg is in transition from gated production plant to urban district, as it has been during the decade that this book explores, namely from 2006 when Carlsberg announced that the site would be transformed from brewery to urban district and to 2016, when the first new buildings opened (0.2, 0.3). In this period, the Carlsberg site was surveyed multiple times in the context of cultural heritage, urban planning and design projects. Especially in the intense planning period between 2006 and 2009 fundamental premises for the ongoing development were created. I will scrutinize the heritage, design and planning acts to ask: *How have heritage professionals, planners and designers understood and assessed the Carlsberg breweries during its urban redevelopment? Which underlying values, assumptions and narratives do these assessments rely upon and how?*

Image 0.3 Prospect of Carlsberg's urban district in a visualization by Entasis Architects from 2009. The 30 ha brewery is planned to become a dense urban district of 3-5 storey city blocks that create narrow streets and trapezoid city squares, combined with high rises that serve as new landmarks when seen from afar. The new district is planned for housing, commerce and cultural institutions.

The official plan for the envisioned city district of Carlsberg follows a particular approach that puts a strong focus on making publicly accessible urban spaces, which are seen as a key value to attracting people to the formerly closed production site. The plan presents urban spaces as a key to successful urban development and gives detailed directions as to how such spaces should be structured to stimulate 'urban life'.[4] Yet, while creating new urban spaces was so key in the redevelopment of Carlsberg, it is striking how the outdoor areas that were already present in the former brewery went largely unseen, despite of the intense site-surveys that were made

4 Entasis Architects, 'Vores Rum,' winning proposal to the international ideas competition 'Carlsberg Vores By' in 2007. The competition entry laid the foundation for the officially approved municipal plan: Copenhagen Municipality, 'Lokalplan 432, Carlsberg II.'

during the redevelopment phase. Outdoor spaces were treated as blank spaces during most of the entire decade of planning that we will study here. Carlsberg's industrial open spaces remained unnoticed, unnamed and their stories untold in heritage surveys and planning documents, thereby slipping explicit discussion during the decision-making. In this period, the brewery is changing by densely built areas, which was an easy operation since little or no value was ascribed to the existing open spaces. I will critically discuss how architects, landscape architects and urban designers as well as planners and heritage professionals contributed to the meaning-making processes during Carlsberg's redevelopment, which was strongly driven by economic motives.

The gap between what got attention during Carlsberg's redevelopment and what did not, between what was deliberately kept and what was removed without much discussion, is an inspirational starting point for this book. It encourages me to add to the understanding of Carlsberg by putting particular emphasis on its post-industrial 'open spaces', which will be used here to denote all outdoor spaces; paved and green, gardens, parking lots, roads, stairs and more, while the broader term 'urban landscape' means all of the physical environment without separating between built and unbuilt or more or less urban or rural parts of the territories. In studying Carlsberg's open spaces, I draw from some of the counter-voices in the redevelopment process, expressed in unbuilt design proposals and heritage surveys that failed to influence the decisions. Some of these did examine the post-industrial open spaces of Carlsberg explicitly and even associated them with certain values worth retaining or strengthening in the future, yet for utterly different reasons. By examining these and by adding on with new counter-stories I want to show that Carlsberg, like any other urban landscape, has multiple potential values. This book thus explores how we can reappraise open space as a widely overlooked, yet constitutive, aspect of Carlsberg as an industrial landscape and seeks to contribute to a more explicit consideration of its characteristics and potentials. I ask: *What characterizes Carlsberg's open spaces and how did they come to be this way?*

This study presumes that industrial landscapes are not indistinct. All aspects of those landscapes – and that includes seemingly anaesthetic or insignificant parking lots and roads – have specific characteristics, shapes and textures. Industrial open spaces did not occur randomly, but reflect certain formation processes, which relate to shifting modes of production, distribution, different social orderings and more. This formation is not only what happened when they were planned, but also what happened after; industrial open spaces may carry distinct social histories of people

Image 0.4 The Carlsberg plan connects much hope with creating good urban spaces, such as this one showing middle class families and other people with an urban lifestyle populating markets and cafés between tall buildings. The image was part of the winning design project for the international ideas competition 'Carlsberg Vores By' 2007, by Entasis Architects.

who have worked in them and shaped them over time, of the adaption to different hydrological conditions and more. It is thus not enough to perceive open spaces in a static way or to merely study their physical characteristics. Rather, I propose to study open spaces as participants in layered social, ecological and cultural processes by drawing on an emerging research tradition called landscape biography.[5] This is not a fixed method, but a

5 The variety of methodologies and research questions in this scholarship is outlined in the anthology *Landscape Biographies: Geographical, Historical and Archaeological Perspectives on the Production and Transmission of Landscapes*, edited by Jan Kolen, Johannes Renes and Rita Hermans, the introduction of which examines the pedigree and key topics in landscape biographical research: Kolen and Renes, 'Landscape Biographies.' My own use of the concept in the study of Carlsberg and its relation to this strain of landscape biographical research has

Image 0.5 The gap between what we give attention and what not, between what we deliberately keep and what we remove without much discussion, is the starting point for this book. Take, for example, the iconic Elephant Gate, built in 1901. This is located at the prominent entrance to the Carlsberg site and is highly valued in the public and professional discussions about Carlsberg.

loose cluster of research activities that has been established in archaeology, cultural history and landscape studies since the 1990s. With landscape biography, we can discover how industrial landscapes 'live' and how they constitute the lifeworld of multiple groups and individuals, as well as how people shape and use landscapes, and interact with them in different ways. Biographical perspectives can play a fruitful double role in studying landscapes that – like Carlsberg – undergo significant change in the present; research can at once provide critical observations of contemporary

previously been outlined in the article 'A Biography for an Emerging Urban District: Discovering Open Spaces in the Former Carlsberg Breweries' in the same book, and elaborated in the first chapter of this book.

Image 0.6 What roles can post-industrial sites have in the future city? Many redevelopent projects seek to inspire new users to connect to former production sites in new ways and welcome playful activities. Picture from the Île de Nantes in France, where people can discover the former ship-building area and its transformation by riding this mechanical elephant that was built for the project.

practices while also contributing with 'depth, nuances or alternative stories' that can fuel new thinking and action in landscapes.[6] This paradoxical, yet potentially productive double-role is the position of this book; at once written by the critical observer and an active participant in the narration and valuation of Carlsberg's living landscape in the context of its ongoing and future reconfiguration.

While previous landscape biographical research has often focussed on rural regions and different urban situations, the potential of such an

6 Roymans et al., 'Landscape Biography as Research Strategy.'

approach towards open spaces is still to be fully explored.[7] To start such an exploration, I will draw on philosopher Henri Lefebvre's notion of space as constantly produced.[8] In his view, space emerges in the interaction of the material, the discursive and the practiced sphere, and this interplay will form the lens through which I engage with Carlsberg as a case study, exploring more closely how this interplay produces industrial open space, starting with the present-day redevelopment process, but also studying how this interplay has worked in the past and can guide alternative futures. The third research question thus concerns the very act of doing this research: *What heuristic strategies can be used to study industrial open spaces in the context of urban redevelopment?*

This book is written out of a strong belief that words and images matter to how we value existing landscapes and thereby also influence how we choose to manage their future changes. To study the past and present of an industrial landscape and to propose scenarios for its future are not separate activities. Rather, studying, knowing and changing are interrelated; a certain reading always implies valuation and a certain action is based on a particular perception of a landscape. Architect and urbanist Manuel de Solà-Morales captures the indivisibility of design projects and site surveys with the following statement: 'To draw is to select, to select is to interpret and to interpret is to propose.'[9] In this interpretative process, stories are a powerful tool, because they are a vital part of how we make sense of the world.[10] I will examine the narratives that were operative in determining Carlsberg's plans and I will provide new, alternative stories that can broaden and deepen the understanding of this landscape. In fact, the act of deepening becomes quite literal at the end of the book, where I examine what is quite physically underneath Carlsberg's open spaces – telling the story of its sub-terrain anew.

The following chapters are not a plea for retaining every bit and piece of former industrial landscapes. The belief here is not that preserving existing open spaces is necessarily good per se, for example, because it strengthens local people's sense of community or of belonging. Neither should the following be understood as an attempt to find an exact model for the city of tomorrow by excavation into the industrial past. Rather, when we imagine

7 An example research using the term 'landscape biography' in relation to industrial sites is Veldhoven, 'Post-Industrial Coal-Mining Landscapes'. In Kolen, Renes, Hermans eds. *Landscape Biographies.*

8 Lefebvre, *The Production of Space.*

9 Solà-Morales, 'The Culture of Description,' 18.

10 Mattingly, 'Narrative Reflections on Practical Actions.'

what the city can become we always relate back to something that is already known; human creativity evolves in dialogue with previous solutions and phenomena. Knowing the physical environment that is actually here better, then – and that includes seemingly insignificant industrial spaces – can thus give us as a society a broader spectrum of possibilities for how we imagine the future – both as to what can be worth retaining and to imagining the aesthetics, uses, meanings and experiences of tomorrow's city.

The issue has become increasingly important during the last five to six decades, when industrial activities of harbours, mines and factories in many European regions and beyond move to locations that are seen as more profitable, leaving the then abandoned sites open for change. Some of the fastest and most encompassing transitions happen when developer-driven urban development projects treat derelict plants as commodities

Image 0.7 Like most other parts of the Carlsberg Brewery that were shaped in the late twentieth century, this space was rarely appreciated and its transformation not explicitly discussed during the urban redevelopment process.
This space was situated on a plateau that gave it a view of the city and an open feel with a lot of sky, as seen here prior to the construction of new buildings on this space.

with new housing, cultural centres, shopping and other urban purposes. The Carlsberg redevelopment was initiated by the global beer company itself, which retained the ownership of the property for the first years of the urban project.[11] Carlsberg Ltd's purpose in doing so was dual: it wanted to sell off the 30 ha/80 acre property piece by piece once built and also to use the future Carlsberg City as a signpost to beer customers and tourists[12] – a signpost in which certain narratives were more wanted than others. I will present new layers and alternatives to the existing storying of Carlsberg, inspired by philosopher and landscape architecture scholar Sébastien Marot's observation that 'a tale cannot be challenged, except with another tale'.[13]

Reconfiguring industrial sites is a complex process involving multiple economic, ecological, political and social challenges. Yet, it also always concerns how people imagine the future, while being situated in the present and, with reference to geographer Donald W. Meinig at the epigraph of this introduction, never free of traces of the past. This cultural dimension is the focus of this book. I will consider traces from the past broadly; as physical remains like buildings, roads and parking lots and as intangible traces like names, practices, and social memories. Such traces are not reflecting a linear development or hierarchical order, but rather exist in a mesh of multiple temporalities that are often referred to as *layered landscapes* – a strong, yet often vague metaphor that is frequently used in design and heritage practice and research.[14] This study explores the temporalities in which Carlsberg's open spaces have emerged, and proposes three ways of understanding landscape change in the concluding chapter.

11 Carlsberg Ltd formerly was the sole owner of the site until 2012, when it was sold to a consortium in which the share of owners were the Charitable Trust Realdania (25%), Carlsberg Ltd (25%), PFA Pension (20%), PenSam (15%) and Topdanmark Insurance Company (15%). In 2017, Realdania left the consortium. See: Development Company Carlsberg Byen, 'Realdaina Sælger sin Andel af Carlsbergbyen,' and Development Company Carlsberg Byen, 'Carlsbergbyen Får Nye Ejere.'

12 Ulvemann Explorative and Carlsberg Ltd Properties, 'En kvalitativ undersøgelse vedr. image af Carlsberg og Apollo.'

13 Marot, *Sub-urbanism and the Art of Memory*, 3.

14 For a discussion of this metaphor, see Renes, 'Layered Landscapes.' For examples of how the term is used in heritage discourses, see Roymans et al., 'Landscape Biography as Research Strategy.' In the design disciplines the concept 'layeredness' has multiple uses, for example, in establishing analytical approaches to landscape that emphasize ecology; see McHarg, *Designing with Nature*, and in urban morphology, see, for instance, Solà-Morales, 'The Culture of Description.'

Carlsberg – A multifaceted case study

Carlsberg's urban project has involved a broad range of international and local experts and its urban plan has received multiple international distinctions.[15] Especially, the international ideas competition 'Carlsberg Vores By' (Carlsberg – Our City) to create a plan for the future Carlsberg redevelopment in 2006-2007 with over 200 design teams from more than 30 different countries provides a rich source into how designers read and value urban landscapes in the context of urban redevelopment. I will use these competition entries together with other design projects that were commissioned before and after and the multiple heritage surveys of Carlsberg, to discuss ways of reading and assessing Carlsberg's industrial landscape.

The competition brief emphasized that the Carlsberg brewery site is recognized as a 'national treasure that contains valuable Danish cultural heritage'.[16] This 'treasured' brewery was to be developed into an urban district, combining housing, commerce and cultural institutions to form what the commissioner envisioned as a 'dense, vibrant and pulsating city district [...] in a setting that is both historical and contemporary'.[17] This outspoken goal of reconciling conservation with development makes the Carlsberg case interesting, especially since it is a complex agglomerate of industrial architecture, some of it dating back to 1847, built on former agricultural properties, but with continuous extensions and reorganization until the 1990s. With this spatial heterogeneity in mind, it is remarkable how the decisions about what parts of Carlsberg might be worth preserving was reduced, quickly and almost without discussion, to a few old, built artefacts and representative gardens. The existing landscape of Carlsberg seems to have escaped the standard lenses that were used for survey and valuation and suggests that we need to develop new ones.

As I began researching the Carlsberg brewery site in 2007, everyone from planners to neighbours and artists were highly fascinated by the visionary urban project that turn the former brewery accessible to the public and turn it into a hotspot in the city's cultural and public life. The Danish public was generally sympathetic towards Carlsberg Ltd and its founding site in

15 The winning entry in the Carlsberg competition (of which the urban plan is a further elaboration and continuation) has won several international prizes (e.g., for the 'World's Best Master Plan' at the World Architecture Festival (WAF), Barcelona, in 2009) and has been included in various international exhibitions, such as 'Green Architecture for the Future' at the Louisiana Museum of Modern Art in 2009.

16 Petersen, 'Invitation', in Carlsberg Ltd Properties, 'Vores By Carlsberg,' 3.

17 Ibid.

Copenhagen was well-known for its remarkable historicist architecture with its rich ornamentation and sculptures of historical personalities and Greek gods. In the early phase of the redevelopment project, its stated aim was to build upon the uniqueness of Carlsberg and to 'reinforce' the 'spirit of the place'.[18]

A decade later the emerging Carlsberg urban district is increasingly criticized in the public, expressing a sense of loss of the former brewery's 'industrial atmosphere', 'edginess' and openness to reuse for public and cultural functions.[19] The financial crisis in 2008 meant that the expected investors in the new Carlsberg urban district did not arrive and the planned construction was postponed. The large industrial halls and open spaces that had been planned for demolition to give way for new buildings were now instead used for exhibitions, concerts, temporary playgrounds and cultural events in what the owner, Carlsberg Ltd, conceived as a waiting time (0.8). Thereby new groups of users – cultural entrepreneurs, young artists and students, families with children – became familiar with those parts of the brewery that had been planned for demolition and began to appreciate exactly those spaces. Yet, decisions about the future of this part of the brewery had been made years before according to economic rationales, but also cultural heritage thinking and urban ideas that this book will unpack.

The Carlsberg redevelopment is in many ways innovative in terms of rethinking how heritage can be defined and actualized in a strategic, urban context. Two strong actors that operated at Carlsberg, the Heritage Agency of Denmark[20] (national authority) and Carlsberg Ltd's property department (the owner of the premises and commissioner of the urban project until 2012 and thereafter co-owner), decided to collaborate closer than ever before.

Rather than fighting over protection versus development as Carlsberg and the heritage authorities had done a decade earlier[21], they attempted to combine the defining of heritage and urban planning and to leave several decisions open until after a vision had been created with the international ideas competition. The Heritage Agency of Denmark explored a new role for itself when it co-wrote the brief for the international ideas competition

18 Carlsberg Ltd Properties, 'Vores By Carlsberg,' 6.

19 Rosenvinge, 'Jeg drikker aldrig mere en Carlsberg'; Holst and Jacobsen, 'Visionen der forsvandt.'

20 During an institutional change in January 2012, the Cultural Agency of Denmark changed its name to the Danish Agency for Culture.

21 Carlsberg Ltd. 'Carlsberg: Fredning er bureaukratisk og erhvervsfjendsk'; Heimann Olsen. 'Carlsberg Bryggerierne imod fredning af sine gamle bryggerier.'

Image 0.8 Carlsberg Ltd attempted to make sure the site became attractive for users, tenants and potential investors despite the pause in planning caused by the recession in 2008. In 2010, along with two foundations and Copenhagen Municipality, Carlsberg commissioned three urban space projects to last for approximately five years.

'Carlsberg Vores By' in 2006, which asked designers to come up with ideas on where and how dense to build, what and how to keep and reuse and what to change.[22] After the competition, Carlsberg Ltd Properties invited the winning designers together with the municipal planners and heritage authorities to a five-day on-site workshop in which they negotiated on how the new urban plan could encounter the existing landscape. Standing at selected sites in the breweries, they negotiated about how specific buildings

22 The brief asked: 'How can the distinct structures in the area be integrated and reused? What rate of utilization would be acceptable in the area?' (Carlsberg Ltd Properties, 'Vores By Carlsberg,' 45).

and gardens should or should not change as part of the conversion.[23] The interdisciplinary collaboration requested the different experts to explain their rationales and concepts much more than many of them were used to from other contexts.

Carlsberg is at once an exceptional and a typical industrial site, but this case study does not claim that it is instrumental to all other industrial landscapes undergoing transformation. Rather, inspired by planning researcher Bent Flyvbjerg, I propose that in-depth case studies are helpful to achieve specific and contextualized knowledge about the 'many-sided' and 'thick' character of each situation.[24] I will present Carlsberg's landscape biography through stories, attempting to make sense of it, but also to keep the possibility open for readers to make sense of it in their way.[25]

Discovering post-industrial open spaces

Some parts of industrial architecture – such as adorned brick factories – are quite likely to be thought of as valuable in contemporary urban redevelopment, while others are more or less intentionally discarded. To me the most valued elements are not always the most interesting to discuss. Rather, I direct attention towards those parts of the industrial urban landscape whose value is characterized by tension, or which go unnoticed in urban redevelopment projects, in particular those industrial facilities that are planned after 1945.[26] Quite typically, the official designations of listing or preservation of Carlsberg's built heritage show how the youngest parts of this landscape were automatically much less likely to be seen as valuable than the oldest ones. The heritage survey focussed on the largest and most adorned buildings by famous architects that were already canonized in historiography and critique (0.9). These criteria seldom applied to the youngest

23 In June 2008, a group of key stakeholders in the redevelopment of the Carlsberg brewery site in Copenhagen spent a week on-site sharing observations and discussing and negotiating reuse and heritage. The group included representatives of Carlsberg (Jacob M. Andersen, Petra Hækkerup, Ulla Nymand), Entasis Architects (Signe Cold), the Heritage Agency of Denmark (Lisbeth Brorsen, Jesper Jensen) and the Copenhagen Municipality (Berit Jørgensen, Helle Hagerup) and Copenhagen Museum (Inger Wiene). I participated as an observer.

24 Flyvbjerg, 'Five Misunderstandings about Case-Study Research', 311.

25 Ibid., 238.

26 In Denmark, buildings younger than 50 years are not assessed as a part of the national heritage unless special exceptions are made. This means that recent buildings are automatically excluded from the discussion of national heritage (Danish Agency for Culture, 'How Does a Building Become Listed?').

Image 0.9 Official heritage designation of Carlsberg. Focus is on buildings, while open space is with few exceptions treated as blank space. The strongest preservation status, listing, is reserved for the oldest buildings or those built by canonized architects. Almost all of the post -1945-parts of the brewery (e.g. the tank parks and large grey buildings to the right) are defined as buildings that may be demolished.

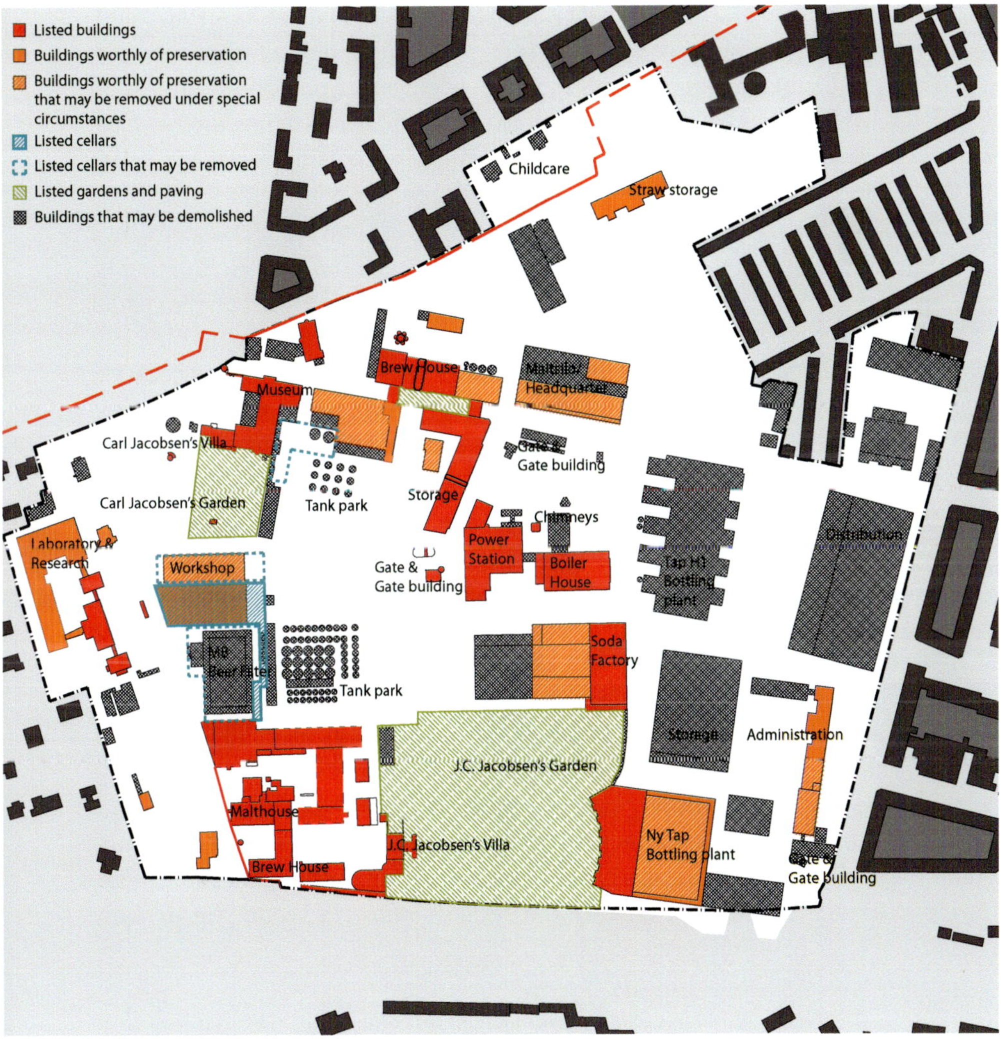

Map from 'Lokalplan 432.' 2009. English names by author.

parts of the brewery and expose a limitation in the existing survey methods. What, then, characterizes industrial facilities built after 1945? Architecture historian Jørgen Sestoft's notion *process complexes* can be useful to show how they are often made for industrial flows of energy, material, people and goods through pipes, streets and other connecting devises.[27]

The Carlsberg breweries were reconfigured numerous times during the last century, becoming a process complex with a network of connecting cellars, pipes and other transportation flows inside and outside of buildings. Carlsberg's process complex, in turn, was connected to larger networks of regional and even global trade, knowledge and production. To assess what is particular about twentieth-century industrial architecture we have to widen the gaze from singular buildings to the larger scale of process complexes and relationships with larger networks. While cultural heritage management has traditionally been oriented around artefacts, open space is an emerging theme in the heritage management of built environments. The Venice Charter from 1964 was a milestone that put focus on the spatial qualities of historical city cores in Europe, many of which were then still war damaged.[28] Many countries have since developed concepts, legislation and survey methods for addressing larger environments and open spaces, such as the German concept *Ensembleschutz* and the Dutch *biotope* used in heritage contexts. In Denmark, the SAVE (Survey of Architectural Values in the Environment) method was introduced in the late 1980s and offered a substantial contribution to addressing the interplay between topography, cultural history and architecture.[29] SAVE was, however, developed in a time when industrial landscapes were not yet high on the agenda in heritage discussions in Denmark.[30] Its methodological breakthrough was crafted in close accordance with those pre-industrial environments that were a big issue then. While open space has emerged as a theme in cultural heritage, the experiences with addressing the outdoor spaces of industry are still cursory, as the Carlsberg case will show.

For the work of architects, urban planners, landscape architects and urban designers, open space has been a much more explicit theme throughout, although in different ways. In the decades following World War II, many

27 Sestoft, *Danmarks Arkitketur*, 92.

28 The Venice Charter was an important milestone in considerations of heritage at an urban scale. See International Council on Monuments and Sites (ICOMOS), 'International Charter for the Conservation and Restoration of Monuments and Sites.'

29 Algreen-Ussing, Silding and De Wal, *Byens Træk*, and *SAVE: Kortlægning og registrering af byers og bygningers bevaringsværdi og udarbejdelse af kommuneatlas.*

30 Jørgensen, 'The Industrial Heritage in Denmark.'

designers argued that urban space had been diminished in modern planning to a degree that it had to be resurrected.[31] They promoted the pre-industrial, classical European town square as the model that could restore urban life; a belief that influenced the Townscape movement, New Urbanism and other attempts to resurrect the qualities of premodern cities.[32] As the following chapters will show, the critique of modernism has been remarkably present in the decision-making around Carlsberg, where the thinking of the most influential Danish critic of modern planning, Jan Gehl, came to make a strong case against the spaces of twentieth-century Carlsberg. Reassessing the industrial open spaces in fruitful ways thus require a critical discussion of this position.

The structure of the book

Chapter 1 ('A Biographical Approach to Industrial Landscapes') situates the recent changes at Carlsberg in the contemporary 'paradigm of transformation' – a condition in which heritage and design practices are becoming increasingly intertwined in the context of spatial planning, while also new ways of conceiving of both heritage and design emerge.[33] These changes, I argue, call for new research that can begin a discussion about industrial open space both in a perspective of the present, the past and potential futures of such spaces. To substantiate contemporary intervention in post-industrial landscapes, the chapter proposes, we need to understand their open spaces particularly and to do so in a dynamic perspective. That is the reason that a biographical approach can bring in vital knowledge, as the chapter argues.

The three following chapters delve into three of the most significant tensions during the urban redevelopment, in terms of how Carlsberg was valued. I use those tensions to raise more general questions about reading and valuing industrial landscapes undergoing transformation.

31 This critique emerges from within the modern movement in the immediate post-war years as well as from professionals who did not see themselves as part of the modern movement; see, for example, Tyrwhitt, Sert and Rogers, *The Heart of the City*; Cullen, *The Concise Townscape*; Jacobs, *The Death and Life of Great American Cities*; and Ellin, *Postmodern Urbanism*.

32 Ellin, *Postmodern Urbanism*.

33 Kolen, 'Rejuvenation of the Heritage'; Janssen, Luiten, Renes and Stegmeijer, 'Heritage as Sector, Factor and Vector'; Braae, Beauty Redeemed; Braae, Riesto, 'As Found'; Riesto, Søberg, Braae, 'City of Open Works.

Chapter 2 ('Site') studies the processes of defining boundaries for the Carlsberg site. I question how Carlsberg was defined as a site – an urban site, an industrial site etc. – during in the redevelopment process, examining how different concepts of unity and diversity, insulation and entanglement were negotiated. Informed by forgotten entries in the design competition and by innovative heritage surveys, I develop new narratives that attempt to nuance and question the standard site definition. The chapter shows that defining site is never self-evident, but rather part of the value-laden choices that are closely tied to decisions in urban redevelopment – both concerning the past, present and future of an industrial landscape.

Chapter 3 ('Space') examines the clash that emerged between the dominating ideas about future urban space and Carlsberg's existing spatial structures. I show how Carlsberg during the first years of the redevelopment became a battleground for critiquing of the modern city, as seen in the open space structure of Carlsberg stemming from the mid- to late twentieth century. The urban planning and heritage survey process also fostered alternative concepts of urban space qualities. Drawing on this urban discussion, I develop new ways of conceiving of Carlsberg's twentieth-century open spaces and their becoming between materiality and imagination, between industrial rationales and the everyday, between soil, water and shifting concepts of beauty.

Chapter 4 ('Sub-terrain') examines Carlsberg's cellars, which came to have a decisive role in Carlsberg's urban plan and its reception. What values did planners, designers and heritage professionals associate with Carlsberg's cellars and why? Informed by the thick discussions that surrounded Carlsberg's cellars I include different voices in the biography of these underground facilities – in new narratives that embrace worker's oral histories and practical jokes, the 19th century novelist Herman Bang's deeply felt visit to the cellars and the role of substances like soil, yeast and ammonia for creating the cellars. The chapter argues that the brewery cellars are just as much technological facilities as they are spaces of human imaginary and it shows how not only contemporary discussions, but also century-old cultural metaphors and words can lurk into our thinking about an former industrial site and ultimately impact on the planning of its future.

The last chapter (Biographical approaches to industrial open spaces') draws general reflections from the study of Carlsberg. I discuss the value of biographical perspectives in terms of surveying and understanding industrial landscapes in the context of urban redevelopment, especially asking about how open space can be understood more closely. I present findings from Carlsberg and reflections about how we can grasp industrial sites as dynamic, yet beyond standard perceptions of linear development. This part discusses who (or what) has changed the landscape of Carlsberg and how. The chapter ends by proposing that landscape change can be understood as (at least) three different types of processes, which emerge from the study of Carlsberg.

1 A biographical approach to industrial landscapes

The starting point for this chapter is the contemporary situation in Europe and North America where focus is more on reusing already urbanized areas than on expanding cities and the challenges and thinking about heritage and urbanism are changing. This situation calls for substantiated knowledge about industrial landscapes and for a reflection on how we intervene in them in the present. A biographical research strategy can be a starting point for developing such knowledge and I will show how this book draws from an emerging cluster of international scholarship around the notion of landscape biography.[1] The end of the chapter shows how I adjust landscape biographical research strategies to a new study area: industrial open spaces during redevelopment, and lays open the investigatory techniques and the sources that I used.

Urban transformation and the need for new heuristic strategies

During the twentieth-century, industrial production plants were built in a spirit of urban expansion while humans increasingly urbanized new territories. Today, such expansion of existing cities is often no longer possible, because the surroundings are already defined by specific functions, be it as roads, agricultural land or built-up space. The modern conception of spatial development as a concentric expansion from the city core outwards to virgin land is increasingly replaced by a recognition that every new urban development implies transforming an existing area. This is tied to a growing awareness that the construction industry is resource hungry and that adapting and reinventing existing areas in cities, including former industrial sites, also involves ecological concerns and a sustainable way of dealing with resources. Adaptive reuse may be easy when it comes to old and spectacular warehouses and brick factories, while other parts of the urban landscape, such as remnants of the industrial period after 1945 are often less likely to be taken care of with the same interest.

1 Some of the arguments in this chapter were published as an earlier version in Riesto 'A Biography for an Emerging Urban District', in Kolen, Renes, Hermans eds. *Landscape Biographies*.

The scientific advisor to the influential International Building Exhibition (IBA) Emscher Park in the German Ruhr district, planner and architect Thomas Sieverts, has stressed the importance of recognizing explicitly the expansive young urban landscapes that make up our environment 'where we live now'[2] – such as suburban shopping malls, train lines, suburban housing districts, industrial sites – which are often perceived as 'nameless' and seemingly indistinct.[3] By acknowledging the characteristics of our physical environment, Sieverts argues, we can expand the understanding of what can be valuable. And if our attitude changes and citizens, users, stakeholders and politicians begin to appreciate more aspects of the environment, we will be encouraged to take better care of it.[4] Informed by Sieverts' thinking, I will excavate those characteristics of industrial landscapes that are underappreciated. The word 'digging' serves as a metaphor for this investigation; it is a long-standing metaphor for empirical research[5] and it suggests that topography will play an important role in the biography of Carlsberg. In colloquial English the word also has another meaning: to make someone aware of something, point towards it, and potentially appreciate it, as in 'dig that!', which is what I attempt to do in the following chapters.

Dealing with abandoned industrial plants entails particular urban planning challenges and I will single out a few of them here. One concerns the relationship that former production sites have with larger urban landscapes. Modern urban planning, as manifested in the Athens Charter, put great efforts into spatially separating residential districts from the 'noise and air pollution 'of industry.[6] Planners systematically separated industrial production from other areas (or zones) in the city, and industrial sites were laid out according to specific rationales and premises that differ from the rest of the city. Yet, industrial sites were simultaneously also strongly entangled in the urban landscape and connected to local, national and, increasingly, global networks of trade, manufacture and economy. This often ambiguous relationship that industrial production sites have with their surroundings is changing once again when on-site industrial activities stop or move. An important topic today is how to inscribe former industrial production sites within the urban landscape in new ways. To

2 Sieverts, 'Where We Live Now.' See also Sieverts, *Zwischenstadt.*

3 Sieverts, 'Where We Live Now,' 21.

4 Sieverts, 'The Prerequisite of a Sustainable City Is Beauty.'

5 Williams, *Notes on the Underground.*

6 Congrès International d'Architecture Moderne (CIAM), 'The Athens Charter.'

address this challenge, a deeper knowledge of the industrial production sites and their participation in different contexts – both in prospect and retrospect – can be vital.

Another challenge is particularly related to those landscapes that were formed by the industry after World War II. While historians, industrial archaeologists and landscape studies scholars have played an important role in showing the strong bonds between landscapes and the industrial activities in Europe since the eighteenth century, much literature does not address with the youngest parts of industrial landscape history, let alone its contemporary changes.[7] Yet, the second half of the twentieth century was one in which the industrial means of production and distribution accelerated and industrial architecture took on a scale larger than ever.[8] Of all industrial buildings in Denmark built in the 125 years that passed between 1840 and 1975, half are from the post-war era 1949 to 1975.[9] The mere quantity of those physical remnants from this young historical period forces us to deal concretely with their characteristics. Also, being a recent past, this is a period to which people who live today connect memories and emotions, and which is closely linked to our present-day societal structures and cultures. There is therefore a particular need to deepen the knowledge about the most recent period of industrial production and to recognize the marks that it has made on the urban landscape.

Young industrial facilities (from the decades following 1945) are often notable for their size and arranged as process-complexes. When investors and future commercial users evaluate the reuse potential of an industrial site, smaller buildings of little economic reuse value as open spaces are considered less of an asset.[10] Open space is a key constituting aspect of industrial process complexes and has often been an important locus for their planning, uses and perceptions. I will therefore examine the role of open space, understood not only as separate spatial units in the outdoors, but as potential participants in flows of the process complex and beyond.

7 See, for instance, Trindler, *The Making of the Industrial Landscape*; Palmer and Neaverson, *Industrial Archaeology*; Hyldtoft, 'Den teknologiske udvikling i Danmark'; Hyldtoft, *Københavns Industrialisering 1840-1914*.

8 Jørgensen, 'Planning the Plant.'

9 Around 43,900 buildings date to the most recent period. See this national survey with GIS-based mapping from 2011: Dansk Bygningsarv and Realdania, *Industriens Bygningsarv*.

10 See for instance, Kasper Lægring Nielsen, 'SAVE som æstetisk og politisk praksis.' In Tietjen, Riesto, Skov eds. *Forankring i forandring*, 162.

Cultural heritage between conservation, transformation and activation of resources

During the last five decades, and especially since the 1990s, the practices and thinking around heritage in the built environment have undergone significant changes. To understand the present situation, it is helpful to recapture what was before, and which Kolen terms the 'grand conservation project'.[11] This describes the thinking when heritage became an established legal practice in European countries during the early nineteenth and into the twentieth century and worked in close connection with nation building. Heritage laws, institutions and later also educational programmes were established to safeguard certain buildings, monuments and objects. These were believed to have a special ability to represent collective values and history as cultural heritage. This urge to safeguard certain objects from change was partly a reaction to the rapid changes caused by industrialization and urbanization and came along with a growing fascination for history. Certain artefacts were defined as cultural heritage and focus was on safeguarding these, and to make sure that they did not change in ways that were thought to disturb their authenticity or capability to represent history. Parallell, but with little concact with this discourse, special legislations and parametres for nature preservation were established.[12] The grand conservation project was characterized by a sense of loss in which, as Kolen and Johannes Renes put it, the 'prime motive was an anxiety over losing valuable objects, buildings, landscapes and traditions – in fact, of the core values of culture itself'.[13]

An important aspect of this heritage paradigm is that the activities of appointing heritage, of registering, conserving and communicating aspects of it belonged to a small elite; it was 'the domain of politicians, learned societies, connoisseurs and later academically trained professionals.'[14] These experts worked relatively detached from the public and from discussions in other domains and often did not identify much need for explicit reasoning and debate. The prevailing assumption was that some buildings and monuments carried an inherent cultural heritage value, although – as Kolen

11 Kolen, 'Rejuvenation of the Heritage,' 50.

12 For contemporary debates showing how this divide is changing, see Harrison, 'Beyond "Natural" and "Cultural" Heritage'; Cameron and Rössler, *Many Voices, One Vision*; and Brockwell, O'Connor and Byrne, *Transcending the Nature-Culture Divide in Cultural Heritage.*

13 Kolen and Renes, 'Landscape Biographies,' 41.

14 Kolen, 'Rejuvenation of the Heritage,' 50. See also Ashworth, Gregory. 'Heritage and the Consumption of Places.' In *Bezeten van Vroeger. Erfgoed, Identiteit en Musealisering*, ed. R. van der Laarse, 193-206. Amsterdam: Het Spinhuis, 2005.

Image 1.1 The IBA Emscher Park (1989-1999) was based on the idea that local heritage values can contribute to the positive development of a location. One of the many projects was the transformation of a former steel works onto the Landschaftspark Duisburg Nord, designed by Peter Latz and opened to the public in 1994.

has pointed out – paradoxically, skilled experts were required in order to 'discover and define this value'.[15]

In recent decades heritage has become more and more intertwined with spatial planning and policies that aim to strengthen planning, policies and other local development perspectives that aim to strengthen the future of a given location. The Internationale Bauausstellung Emscher Park (IBA Emscher Park 1989-1999) in the Ruhr region of Germany is a central point of reference for such processes. The IBA Emscher Park built on the idea that local heritage values can positively contribute to the desired development of a place; in this case an entire region that had lost its vital role as the

15 Kolen, 'De biografie van het landschap,' 50. It is important to note that there are many nuances to the overall story here told. See, for instance, art historian Alois Riegl in his 1903 text 'Der moderne Denkmalkultus. Sein Wesen und seine Entstehung.'

coal-and-steel heartland of Germany and which was characterized by high unemployment, decline and a loss of future prospects when the project began. IBA bundled numerous projects spanning from detoxification of water to communication, rediscovery and appropriation of the local urban landscape to participatory planning and large iconic art-and-design projects like the Landschaftspark Duisburg Nord by landscape architecture firm Peter Latz (1.1).[16] IBA not only involved activating and restoring existing buildings and iconic cranes, but also positively redefined other aspects, such as the 'industrial nature' and 'post-industrial landscape', which became shared notions in the public realm. What was applied was not the logic of the 'grand conservation project', but rather a perspective of asking how heritage can contribute to creating a desired local development. According to this thinking industrial utilitarian structures, tracks and traces of the everyday life in the mines and even the flora and fauna in this production landscape were defined as local values that could be a stepping stone towards new prospects in the region. Since the developments in the Ruhr region, local projects across Europe demonstrate this ongoing conflation of heritage and spatial development. This new role and way of approaching heritage has been termed the 'paradigm of transformation' by Kolen[17] and is related to discussions about the redefinition of heritage called the 'New Heritage' by Cornelius Holtorf and Graham Fairclough.[18]

Characteristic of this transformation paradigm is the stance that the objective of heritage is not just to protect representatives of the past and to narrate authorized histories, but, above all, to promote desired local political, economic, social and spatial developments.[19] The European Council's Convention on the Value of Cultural Heritage for Society testifies this thinking and states that in the context of local development heritage value is no longer reserved for an exclusive collection of special artefacts as opposed to non-heritage objects, but rather potentially 'includes all aspects of the environment, resulting from the interaction of people and places through time.'[20] This dismisses previous dichotomies, such as those manifested in practices that distinguished between cultural and natural heritage, young

16 Storm, *Hope and Rust*, and Braae, *Beauty Redeemed*.

17 Kolen, 'Rejuvenation of the Heritage.'

18 Holtorf and Fairclough, 'The New Heritage and Re-shapings of the Past'; Fairclough, 'New Heritage Frontiers.' See also Paludan-Müller, Carsten. 'Kulturarv: Forskning og forvaltning. Generelle tendenser i Europa.'

19 See, for example, Netherlands State Government, *Nota Belvedere*, and *Kulturarven – et aktiv*.

20 Council of Europe, 'Council of Europe Framework Convention on the Value of Cultural Heritage for Society,' 2.

and old history, fine architecture and everyday buildings and makes landscape an important common ground for heritage discussions.

What makes some aspects of reality particularly valuable as heritage, then, in heritage acts characterized by the transformation paradigm, is when a collective of people attach meanings and a sense of ownership to them. Cultural heritage, the Faro Convention states, is 'a group of resources inherited from the past which people identify, independently of ownership, as a reflection and expression of their constantly evolving beliefs, knowledge and traditions'.[21] The continuity implied in the sentence – that a group of people inherit something and identify with it – is difficult when it comes to former industrial sites that are being converted by urban projects; former employees leave and new users enter, and the question of who is 'local' is very explicitly a matter of point of view and debate; In the case of Carlsberg the 'locals' in the period of my research changed from workers carrying generations of oral memories, to new cultural entrepreneurs, gallerists and skaters, developers who have had their offices on-site while working on the Carlsberg City project, and – in the last year of my research – inhabitants in and users of the newly built high-rises. These groups and individuals have utterly different viewpoints on Carlsberg's potential heritage values and the process of inheriting a collective belief is conflicted and complex.

The Faro Convention commits to link cultural heritage to 'the construction of a peaceful and democratic society [...] and the promotion of cultural diversity'.[22] In the transformation paradigm new possibilities arise to address the conflicts and dissonances that are inevitable to any heritage process.[23] This involves attempting to incorporate 'greater synergy of competencies among all the public, institutional and private actors concerned'[24] and pilot new methods of collaboration and participation in multiple heritage processes in different countries.[25] With the emerging transformation paradigm, the question about what may be the public value of heritage is becoming a multifaceted debate[26] and heritage is connected to rationales of social, economic and ecological sustainability.[27]

21 Ibid.

22 Ibid.

23 Tunbridge and Ashworth, *Dissonant Heritage*.

24 Council of Europe, 'Council of Europe Framework Convention on the Value of Cultural Heritage for Society,' 2.

25 See, for instance, Riesto and Tietjen, 'Doing Heritage Together'.

26 Guttormsen and Swensen, *Heritage, Democracy and the Public*.

27 Fairclough, 'New Heritage Frontiers,' 35.

Image 1.2 Who are the locals of an industrial redevelopment site? Picture of tanker drivers emplyed at Carlsberg. The photo was taken in the summer 2008, while many production activities had already moved and just before the site would open to the public.

The transformation paradigm is part of a cultural era where concepts and thinking from the economic domain are increasingly adopted in other fields. Words like 'asset,' 'value,' 'resource' and 'experience economy' frequently appear in discussions about cultural heritage in the built environment.[28] This tendency can be studied by looking at how the Heritage Agency of Denmark acts and presents itself in the years that Carlsberg's planning went on, for example, in 2006, when the agency commissioned a report that estimate the economic benefit of cultural heritage work for municipalities.[29] During the Danish Year of Industry (2007)[30], the agency stressed their desire for increased collaborations with 'private investors and other actors in the public, as well as citizens' (notably mentioned in this order).[31] Explicitly dealing with cultural heritage has become a way for urban projects to distinguish themselves in the attempt to attract tourists, residents and investors. The power of corporate storytelling in urban projects can be studied at Carlsberg, where the urban project encouraged a heroic corporate history, which was countered by other voices in the process.

The transformation paradigm should not be mistaken as a situation where a 'new' heritage paradigm is simply replacing what consequently may be called the 'old' one. Rather, this is a dynamic field of co-existing paradigms that allow the actors that are involved in heritage processes to apply various concepts on heritage, depending on the context and aim of the heritage decision.[32] The relationship between these heritage paradigms can, however, also be seen as a discourse in which different viewpoints contest each other. In her book *Uses of Heritage*, heritage theorist Laurajane Smith identifies an especially powerful and hegemonic contemporary discourse that is practiced by the (often national) heritage authorities. This 'authorized heritage discourse', which carries with it the thinking of the grand conservation project outlined above, Smith argues, 'reinforces the idea of innate cultural value tied to time depth, monumentality, expert knowledge and aesthetics'.[33] This thinking, in turn encourages a hierarchy in which those parts of the built environment that are considered 'grand, old and beautiful'[34] score highest. This discourse is still strong because it is embedded in

28 *Kulturarv*. See also Dümke and Gnedovsky, 'Social and Economic Value of Cultural Heritage.'

29 *Kulturarv*.

30 Jørgensen, 'Industrisamfundets kulturarv – Kulturarvsstyrelsens første satsningsområde.'

31 Hvass, 'Forord'. In Jørgensen, ed. *Industri Industri*, 4.

32 Janssen, Luiten, Renes, Stegmeijer, 'Heritage as Sector, Factor and Vector.'

33 Smith, *Uses of Heritage*, 299.

34 Ibid., 11.

legal frameworks, methods of inventory and established practices of survey and preservation. We will investigate its workings and contestations in the Carlsberg redevelopment process, where the Heritage Agency of Denmark partly enacted an 'authorized heritage discourse' position.

This outline of different heritage paradigms testifies that heritage is never just a matter of unveiling static values in objects – but neither is it merely belonging to human cognition and social domains. The material world is relevant to how we think and do heritage – for instance, the present-day expansive landscapes of industry fuel us to adjust our thinking and actions. To grasp heritage as something that concerns the human world, but also is connected to the flora, fauna and material world, I conceive of heritage as interplay between landscapes and the people who do heritage. Archaeologist Rodney Harrison offers a useful scope in defining heritage as 'a material discursive process in which past and future arise out of dialogue and encounter between multiple embodied subjects in (and with) the present'.[35] Heritage, then, is both the processes and the products of these process.[36]

Urbanism and three ways to define place

In the fields of planning and design significant changes have taken place during the last five to six decades and it is useful to recall how modernism and its critique as a backdrop for European urbanism. At the beginning of the twentieth century a prevailing modernist belief was that designers and planners could work free of constraints from the past in an evolutionary development towards a better future. Le Corbusier's Plan Voisin was highly influential because it defined architecture as a tool to create new living conditions and a better society through massive physical master plans. The plan proposed to demolish large historical areas in central Paris and insert a new future city with little sign of interest for the existing physicality on-site. Voisin was never realized, but numerous other modernist plans with a tabula rasa approach to the existing environment were. I should already note that this short outline does not cover all the nuances and exceptions that should also be taken into account; many modernist architects were much more aware of the geographical contexts of their work than they are often thought to be.

35 Harrison, 'Beyond "Natural" and "Cultural" Heritage,' 28.
36 Fairclough, 'New Heritage Frontiers'; Riesto and Tietjen, 'Doing Heritage Together.'

The hopeful tabula rasa planning was quickly challenged in the post-war era of bombed city cores across Europe and collapsed grand ideologies. Many were concerned with what they perceived as a loss of connection to history and expressed a need to depart from utopian ideas.[37] At post-war meetings of the influential group of international modernist architects and planners Congrès International d'Architecture Moderne (CIAM), designers gathered with a renewed interest in the existing city fabric, as, for example, the British designers of the Independent Group's approaches the city with respect for the everyday situation and materials 'as found'.[38]

Next to such attempts to renew modern architecture and planning from within, there was also a more fundamental request to invent alternatives from a growing number of people who criticized the built results of modern planning. The modernist urban fabric was attacked for lacking the ability to encourage social encounters, promote well-being and fulfil a basic human need for meaning. Architects, planners, historians, psychologists and sociologists began to search for new urban models by investigating the historical architecture, especially the European historical city centres – many of which had been seriously damaged during World War II. Out of this debate grew concepts such as 'contextualism', '(critical) regionalism', and 'townscape'.[39] Frequently underlying these concepts were essentialist beliefs that certain historical building types met the human need for identification and that certain *places* – a concept that was very much used in this debate – could bestowed a special identity, character or spirit through architecture.[40]

One of the influential critics of modern urbanism was architecture historian Christian Norberg-Schulz, who is especially known for emphasizing the importance of landscape. In his book *Genius Loci*, Norberg-Schulz argued that the role of the designer was to mimic or to complement what he called *genius loci*, a Latin term that means the '(guardian) spirit of a place'.[41] The role of the designer, according to Norberg-Schulz, was to desipher the spirit of the place and then engage with it by adding certain architectural forms. Norberg-Schulz writes from the premise that an (almost static and certainly undisputed) spirit of a place can be revealed by skilled experts and that

37 Heynen and De Jonge, *Back from Utopia*.

38 Lichtenstein and Schregenberger, *As Found*.

39 Ellin, *Postmodern Urbanism*, and Nielsen, *Gode Intentioner*. See also Forty, *Words and Buildings*.

40 Nielsen, *Gode Intentioner*, 77-99, and Ellin, *Postmodern Urbanism*.

41 Norberg-Schulz, *Genius Loci*.

this can be done by relying on visual perception.[42] His idea of *genius loci* has become heavily contested because it fails to acknowledge that places and cultures change over time and that there is not one spirit, but multiple perceptions of the same place, which in itself is not a stable entity.[43] Yet, despite of this scholarly critique, the notion of the *genius loci* is still going strong in part of the public debate about architecture and urbanism. This book will study how the idea of a spirit of place became a powerful actor during the redevelopment of Carlsberg, especially how it came to influence the decisions made about its new urban squares.

Norberg-Schulz was one of the architectural voices in a larger scholarly discussion in the 1970s and 1980s where thinkers from different disciplines began to engage with the concept of place much more strongly than before. Norberg-Schulz did so as a reaction to what he thought was a lack of sensitivity towards place in modern planning; scholars from human geography and other disciplines reacted towards prevailing ways of conceptualizing and analysing landscapes and developed new humanities-based frameworks for discovering place. With reference to geographer Tim Cresswell, I call these approaches the 'humanistic discovery of place'.[44] One of the most influential voices in the post-war place geographical discourse on place is Edward Relph. In *Place and Placelessness*, Relph defines places as 'significant centres of our immediate experiences of the world'.[45] Like Norberg-Schulz, Relph drew from Martin Heidegger's writings about *Dasein* as a basic human condition – that experience is always already situated in the world in a form of being-in-the world.[46] Relph forwarded the idea that being in place is an inescapable human condition, and that places should therefore play a key role in how we understand landscapes – not as neutral geographic information system (GIS) data or as topographical entities or abstract spaces that could be described objectively or from a distance.[47] Rather, Relph and other writers focus on the existential relationship between people and place and into the bodily and hermeneutic dimensions of our understanding of places. What is less discussed in his research, as also in Norberg-Schulz' architectural theories, is how places are contested and involve conflicts of interest and power.

42 Ibid.; see also Riesto, '(Hvordan) skal vi huske Christian Norberg-Schulz?,18-23.
43 Hvattum, 'Stedets Tyranni'; Haddad, 'Christian Norberg-Schulz.'
44 Cresswell, *Place*, 18-24.
45 Relph, *Place and Placelessness*, 141.
46 Cresswell, *Place*, 18-24; Norberg-Schulz *Genius Loci*, 10.
47 Cresswell, *Place*, 25; see also Seamon and Sowers, 'Place and Placelessness.'

The dissonant aspects of place were put forward in a relevant critique from many perspectives; Feminists (like bell hooks) show how the discussion about place had been centred around male experiences of place, while Marxists stress the economic processes and distribution of power to which place contributes.[48] Most famously in this critique, geographer David Harvey said: 'Place, in whatever guise, is [...] a social construct.'[49] Harvey provokingly suggested that the 'only interesting question that can then be asked is: by what social process(es) is place constructed?'[50] One of the strengths of such a radical constructivist position is that it enables us to see that places change and may thus be changed deliberately towards the better; in Harvey's view working towards more inclusive and just power relations. Social constructivism cannot, however, help us understand the relationship between the physical environment and social and cultural aspects.

Although not sharing Harvey's Marxist agenda, a strong voice in urbanism, architect and theorist Rem Koolhaas, has also been engaged with countering essentialist place concepts, yet concluding with another call for action than Harvey did. Koolhaas' widely read essay 'The Generic City' from 1995 presented the entire idea of place identity as a 'mousetrap'[51] that 'restricts' people because it 'resists [...] renewal and contradiction'.[52] His argument takes its starting point in the often superficial marketing of city centres that has come to follow with much urban renewal (at the cost of e.g. suburbs and other less touristic and symbolic areas). Further, the attempt to strengthen the identities of city centres, according to Koolhaas, just ends up being superficial; Central Paris, for example, in its attempt to market itself as even more Parisian, ending as a mere 'hyper-Paris'.[53] The relentless 'routine', Koolhaas continues, 'to transform housing into offices, warehouses into lofts, abandoned churches into nightclubs, [...] precincts to the relentless conversion of utilitarian space into 'public' space(...) all authenticity is relentlessly evacuated.[54] Quite parallel to Koolhaas' writings, social scientists, such as sociologist Sharon Zukin, examined the processes in which renewal and gentrification take away the

48 Cresswell, *Place*, 25-27; hooks, *Yearning*; Harvey, *Justice, Nature and the Geography of Difference*.

49 Harvey, *Justice, Nature and the Geography of Difference*, 261.

50 Ibid.

51 Koolhaas, 'Generic City,' 1248.

52 Ibid.

53 Ibid., 1249.

54 Ibid., 1249.

'authenticity' of places.[55] Koolhaas in his usual provocateur style, on the contrary, affirms the generic city – such as uniform business districts as 'liberated' from the 'straightjacket of identity'.[56] Place identity, he argues, should have no relevance for cities in the future.[57]

As a reaction to those poles of humanistic (and potentially essentialist) and sociopolitical (and potentially social constructivist) ways of understanding place, many nuanced positions exist.[58] For example, geographers such as Robert Sack and Jeff Malpas, conceive of place as 'a phenomenon that brings these worlds together'.[59] In recent years the discussions about place has broadened in multiple directions which can be bundled with the notion *relational*, which to me means dealing with relations in at least three ways: First, relational perspectives have been applied to deepen the understanding of the relationships between *practices and landscapes*, such as in the writings of theorists like Michel de Certeau and Henri Lefebvre, both providing close readings of everyday practices in a way that, to quote Cresswell, sees place as an 'event rather than ontological thing'.[60] Second, the relational is apparent among those many contemporary writers who stress the flows, connections and, *unboundedness* of places. Geographer Doreen Massey is one of the most influential scholars in this strain of research, as well as urban design theorists Carol Burns and Andrea Kahn.[61] A third way of relational site thinking is to examine places through the *dynamic interaction between social, cultural and environmental processes*, which, for instance, geographer Matthew Gandy does in his study of the emergence of New York City as an interaction of city and nature and William Cronon in his study of how Chicago emerged as 'nature's metropolis'.[62] These three types of relational perspectives contribute to deepening and widening the understanding of places in different ways and can be useful perspectives in the study of industrial landscapes undergoing redevelopment.

In the design discourse scholars are also increasingly forwarding relational perspectives as to understand the relationship between the

55 Zukin, *Naked City.*

56 Koolhaas, 'Generic City,' 1249-1250.

57 Ibid.

58 Cresswell, *Place*, 30.

59 Ibid., 31. Malpas, Jeff. *Place and Experience: A Philosophical Topography.* Cambridge: Cambridge University Press, 1999.

60 Ibid., 39. Certeau, *The Practice of Everyday Life*; Lefebvre, *The Production of Space.*

61 Massey, 'A Global Sense of Place'; Burns and Kahn, *Site Matters.*

62 Cresswell, *Place*, 43; Gandy, *Concrete and Clay*; and Cronon, *Nature's Metropolis.*

existing site and the designer who proposes intervention. Landscape architecture theorist Sébastien Marot, for instance, encourages designers to engage thoroughly with the existing site it in their plans in order to 'deepen the territories'[63] in which we live. His plea is a direct reaction against Koolhaas' ideas about how place identity is nothing but a construction that designers need not worry about. Yet, Marot does not think of places as fixed entities that designers should neutrally reveal, as in the essentialist take of Norberg-Schulz. Rather, Marot presents sites as multi-scalar parts of a 'dialectical landscape' which 'is physically and continually undertaken' in the interaction between people and site 'and not merely contemplated as a thing in itself'.[64] The role of designers and planners, then, is neither to reveal an inherent spirit nor to just go with the flow of generic cities, but to contribute to the dialectics of landscape by deep and 'inventive analysis.'[65] Site survey then becomes an interpretative activity that engages with existing aspects, cultures and memories of landscapes, without yet thinking that these are immutable and objective.

Such a conception of design as an activity that involves the creative 'reading' of landscapes is becoming increasingly relevant in the transformation paradigm, when the main urban task is to contribute to 'editing' places that are already built.[66] Marot is part of a larger move in research to conceptualize design as *intervention* in existing situations – as action that alters and participates in their ongoing evolving processes – rather than as creating the new.[67] An inventive analysis provides an interpretation of an urban landscape, which in turn can be the subject of a critical discussion. Further, design projects are the result of dialectical engagement with a landscape and can lead to new knowledge about the landscape.[68] This twofold character of design projects – as discursive entities and knowledge production – will be tested using design projects as sources to better understand Carlsberg. Heritage proposals have the same double role in this study.

63 Marot, 'Sub-urbanism/super-urbanism,' 21.

64 Ibid., 23. Marot derives the term 'dialectical landscape' from artist Robert Smithson.

65 Marot, 'The Reclaiming of Sites,' 49.

66 Braae and Diedrich, 'Site-Specificity in Contemporary Large-Scale Harbour Transformation Projects.'

67 Braae, *Beauty Redeemed*; Tietjen, *Towards an Urbanism of Entanglement.*

68 On the discourse of design as knowledge production, see Prominski, *Landschaft entwerfen*, and Seggern, Werner and Grosse-Bächle, eds. *Creating Knowledge.*

A biographical approach to landscapes

Spatial planning, design and cultural heritage deal with the role of traces from the past in the future city, yet as stated by Nico Roymans et al. professionals in these fields have often 'not adequately acknowledged the dynamic way in which the [...] past is perceived and dealt with'.[69] Assessing buildings or places is not an isolated activity, but is gradually built up over time. What others have said and written will consciously or unconsciously influence how we perceive of a landscape and decisions about protection, demolition, alteration and reinterpretation. As urban historians Aidan While and Michael Short write about redevelopment, it builds upon judgements that 'are situated within a set of frameworks, discourses and collective understandings of heritage/design identity of a particular locale enacted at different levels of decision making and accumulated over time'.[70] A better understanding of such accumulated meanings and of how industrial urban landscapes are interpreted and reinterpreted, shaped and reshaped, can provide vital knowledge for addressing former industrial landscapes in redevelopment. Further, in the transformation paradigm where the conditions, thinking and practices of heritage, design and planning are mutating, an increased knowledge migration between different fields and between different positions within these fields, should be strengthened. To contribute to enriching this discussion, I will draw on research around the notion landscape biography.

The word 'biography' derives from the Greek *bios* (life) and *graphein* (writing, describing) and refers to written accounts of a person's life. Biographies describe someone's life over a relatively long time span; that element of time is an important reason that the term 'biography' has travelled to other fields than the book genre. Biography has been applied to diachronic investigations in anthropology, cultural, architectural and urban history for decades.[71] I will here relate to what biography means in relation to studying landscape, as it has been increasingly applied by archaeologists, cultural historians and human geographers since the 1990s.[72]

Biographical approaches to landscape have been quite visible in the Netherlands in recent decades; the notion has been developed in projects

69 Roymans et al., 'Revitalizing History,' 404.

70 While and Short, 'Place Narratives and Heritage Management,' 5.

71 See, for example, Kopytoff, 'The Cultural Biography of Things'; Appadurai, 'Introduction'; and Samuels, 'The Biography of Landscape.'

72 See, for instance, Kolen, 'Recreating (in) Nature'; Roymans, 'The Cultural Biography of Urnfields'; Gerritsen, 'To Build and to Abandon.'

by national funding programmes and connected to policies that aimed at increasing the role of cultural history in spatial planning.[73] An eight-year interdisciplinary research project explored biographical approaches to landscapes, and the 2010 article 'Landscape Biography as Research Strategy: The Case of the South Netherlands Project,' by Nico Roymans, Fokke Gerritsen, Cor Van der Heijden, Koos Bosma, and Jan Kolen, served as a preliminary conclusion. It discusses how biographical approaches to research can contribute to practices of spatial planning and heritage management.[74] In 2016, Jan Kolen, Johannes Renes and Rita Hermans edited the second significant gathering of research in this emerging field: the anthology *Landscape Biographies: Geographical, Historical and Archaeological Perspectives on the Production and Transmission of Landscapes.*[75] With contributions by researchers from a broad range of disciplines and universities across Europe, this collection shows the wide span of approaches that can be gathered under the umbrella of landscape biography. Its introduction, written by Kolen and Renes, reflects on the philosophical foundations of landscape biographies and situates them in the scholarly field of landscape studies. This article and this book provide the basis for my understanding of landscape biography.

Landscape biography is not a fixed methodology, but, as geographer Hayden Lorimer writes in the preface to *Landscape Biographies*, a 'lens through which to view changes made to the natural or built environment, to its social purpose or cultural imagination'.[76] This lens is indebted to phenomenological thinking and understands landscapes as essentially *human lifeworlds*[77]; it is a fundamental human condition that we inhabit landscapes and can never view them objectively from an outside position. Kolen and Renes state that 'being shaped *by* the world, and presence and realization *within* and *through* the world, takes place in a dialectical movement'.[78] Biographical perspectives can help us investigate this dialectical movement with its hermeneutic processes where people interpret and reinterpret the landscapes that they are in, with and shaped by. The metaphor biography is thus double: it describes the life of individuals in landscapes, but it also narrates the 'life histories' of landscapes themselves, evolving through long

73 Netherlands State Government, *Nota Belvedere*.

74 Roymans et al., 'Landscape Biography as Research Strategy.'

75 Kolen and Renes, 'Landscape Biographies,' 21-43.

76 Hayden Lorimer, in Kolen, Renes, Hermans, *Landscape Biographies*, p. 17

77 Kolen and Renes, 'Landscape Biographies,' see, e.g., p. 25, where they lay open how geographer Marwyn Samuels used the concept landscape biography.

78 Ibid., 31.

time spans with shifting cultures and inhabitants. Studying landscapes from a biographical perspective enables us to explore how social groups and individuals use, interpret and reinterpret landscapes, negotiate different meanings of which many accumulate over time.

Having said that landscape biographical studies stress cultural dimensions, however, Kolen and Renes also recognize that landscapes are not confined to the lifeworlds of humans.[79] Rather, as they write '[a]nimals, other organisms, maybe even "animated" objects and places (in specific cultural perceptions) or human-made technologies – all belong to the group of cohabiting actors that together make the landscape into a life world'.[80] Such an expanded notion of lifeworld seems suitable to describe industrial urban landscapes, which emerged in the interaction between rivers, coasts, wind and other landscape phenomena that could be utilized in industrial production. Exactly this interaction between people and landscape is important to understanding industrial urban landscapes, and here biographical studies give an opportunity to view 'landscape at each point in time as the interim outcome of a long-standing and complex interplay between agency, structure and process'.[81]

Such interplay often slips away if we stay strictly within the boundaries of expert disciplines and compartmentalize aspects of landscapes such as soil development, building styles or botany. To apply a biographical lens on landscape requires one to transgress such specialist domains to get a better understanding of the relationship between the physical landscape and the way that we inhabit and understand it or, as Kolen writes, 'between the ecological and social dimensions, and between functions and cultural meanings'.[82] In this way the lens of biography needs in fact to be a combination of multiple lenses. This can be done in collective interdisciplinary research projects, but also in a book like this, where I build a kaleidoscope to see Carlsberg from multiple angles. This, again, has requested me to use multiple sources and investigatory techniques. Such a study will have to lose some of the benefits of in-depth, specialized research from one disciplinary perspective. Yet it can create another kind of knowledge, which can be fruitful to better understand Carlsberg's open spaces in particular, but also processes in which industrial landscapes are reconfigurated more general.

79 Ibid., 32 with reference to Ingold 1993, 2000.
80 Ibid., 32.
81 Ibid., 28.
82 Kolen, 'De biografie van het landschap,' 297.

While many strains of research apply multi-method approaches, what is particular to biographical reserach strategies is that they attempt to transgress the strict distinctions between the present and the past, and even potential futures. Biographical lenses can grasp how people have related to a landscape over time and also form the basis for a critical examination of *present-day evaluations* conducted during redevelopment processes and landscape management. Roymans et al. write that landscape biography 'runs from the later prehistory up to the present-day, and (...) focuses on the study of the interrelationships between spatial transformations, social and economic changes and the construction of regional and local identities'.[83]

By focussing on meaning-making processes, interpretation and narration of landscapes, biographical studies can potentially contribute with critical insights into the management and transformation that is going on in the present. Such a contemporary perspective opens an opportunity to critically examine the present redevelopment at Carlsberg, including a critical analysis of the positions taken in the recent redevelopment of Carlsberg from which designers, heritage managers, the commissioner (Carlsberg Ltd), journalists, users and Copenhagen Municipality associated Carlsberg with values and ideas. Yet biography is not solely a critical endeavour. Rather, as I will test in this book, it can include alternative histories that enrich the understanding of the present landscapes and the scope of possibilities for imagining its future.[84]

Combining the contemporary with a historical dimension opens up the possibility to understand the role of inherited myths and long-standing landscape processes and to study redevelopment projects as part of longer cultural processes. Biographical perspectives have primarily addressed large rural landscapes and archaeological sites, while the contemporary urban space design and urban redevelopment are less explored. My endeavour here is to move the lens of landscape biography over to the field of design for a while, zooming in to how these professionals craft different interpretations of landscape in negotiation with other actors.

Landscape biography has not been developed as a 'pure' academic activity. Rather, the aim has been to dissolve the traditional separation between history, 'often defined as an intellectual interest in the past per se'[85] and planning, heritage management and other practices and policies. Landscape biography is thus both an embedded and a critical research strategy – a

83 Roymans et al., 'Landscape Biography as Research Strategy,' 337.

84 Ibid., 355.

85 Ibid., 337.

paradox that is not without difficulty, but can nevertheless enrich our thinking and actions in paradigm of transformation, so I will walk the line.

Temporalities of landscapes

While the biographical research tradition explicitly studies the present interpretation, management and future-making of landscapes, it also stresses the processes of the past, including longue durée perspectives.[86] Long historical time spans can enrich the understanding of industrial landscapes, which are not confined to notions of the industrial period. Periodization is a productive analytical construct, but does not produce containers of separate pasts. So when, for instance, a new heritage category was introduced in Denmark during the Year of Industry in 2006, called 'Industrial Sites of National Significance', it contributed to a relevant appraisal of industrial heritage and fuelled important new studies of industrial sites and their relationship to technological and societal developments in industrial society.[87] The limitation of such specialized examination of one period is that it can downplay aspects that are not related to industrial rationales and that are older than the industrial period. Here, the temporal scope of biographical studies – spanning from longue durée to future imaginaries – can add to the existing knowledge. Further, if we study landscapes as multidimensional lifeworlds in such a long-term perspective, we cannot – or at least not only – apply a linear understanding of time. We must acknowledge that imprints from different periods intertwine in ways that involve rhythm, repetition, resurrection and persistence in the times of individual people and landscapes.

The term 'layered landscape' grasps some of this complexity and is widely used in various strands of practice and research. Reynolds et al. propose that the concept layered is useful to show two things: first, that landscapes have both diachronic and synchronic dimensions. And secondly, that the past has a 'multi-vocal character (...) in which different groups and individuals raise very diverse questions'.[88] Yet, despite of the power of this metaphor, it is also problematic.[89] How interconnected are such layers? Does this metaphor hint at an ordered layering of time as plates in a stack?

86 Braudel, 'Histoire et Sciences sociales.' See the discussion about the uses of this concept in Kolen and Renes, 'Landscape Biographies,' 28.

87 See, for example, Selmer, *Industrisamfundets Kulturarv*.

88 Roymans et al., 'Landscape Biography as Research Strategy.'

89 Renes, 'Layered Landscapes'. In Kolen, Renes, Hermans, eds. *Landscape Biographies*.

Reflecting on landscape temporalities is one of the key points in landscape biographical research.[90] A possible point of contribution, Kolen and Renes argue, is to conduct that landscape historians can 'conduct more research on the interactions between mnemonic systems, values and perceived time on the one hand, and environmental rhythms and ecological changes on the other'.[91] Studying Carlsberg's landscape can potentially enrich such an understanding of layered landscape change – something I will return to it in the last chapter.

The metaphor biography also has problematic implications. It raises expectations that the research will be presented in a comprehensive story about someone, or something, who is born, matures and then passes away. Such a teleological conception of history cannot be applied to landscapes without great reduction. Yet, literature theory opens up alternative ways to understand what a biography may be. Literature theoretician Mikhail Bakhtin argued that any book always participates in social contexts of heteroglossia (many voices that tell different stories), There are often several characters involved and the author uses words whose etymology is imprinted by those who have used it before him or her. Bakhtin says that cultural utterances participate in such infinite operations and always gather 'specific world views, each characterized by its own objects, meanings and values. As such they all may be juxtaposed to one another, mutually supplement one another, contradict one another and be interrelated dialogically.'[92] With a heteroglossic understanding of books, a biography of Carlsberg need not describe a comprehensive chronological story, but many narratives. This is what I aim to reflect by structuring this book in thematic sections, which combine synchronic studies of the present debate with diachronic ones that narrate multiple stories about Carlsberg's becoming.

Studying open spaces with Lefebvre

To be able to study industrial open spaces, we need a refined understanding of what space is, beyond pure aesthetical perspectives of architectural form, and also beyond isolated sociopolitical understandings that forget the material dimension. A writer that has taken on the task of developing a concept that can grasp space in its multiplicity is philosopher Henri Lefebvre, whose book *La production de l'espace* from 1974, appeared in

90 Kolen and Renes, 'Landscape Biographies,' 38-41.

91 Ibid., 40.

92 Bakhtin, *The Dialogic Imagination*, 291-292.

an English translation in 1991.[93] Lefebvre has contributed to a shift in the focus on the thinking about space as stasis to how it is produced in infinite processes.[94] This focus on process provides a useful starting point to understand such a dynamic topic as industrial open space during transformation.

Another influential contribution in his theory is that it allows us to transgress the binary thinking that would single out the dimensions of space as either objective/subjective, material/mental or real/imagined.[95] Instead, Lefebvre built a theory to understand spatial production in the interface between that which we can perceive (*le perçu*), and that which we conceive (*le conçu*), spaces as invented and imagined, thought of, spoken about, represented and discussed. In this way, Lefebvre combines materiality with discourse, but importantly he also introduces a third dimension that is no less important – our lived experiences (*le vécu*). By combining the perceived, the conceived and the directly lived Lefebvre aims to 'reconnect elements that have been separated and to replace confusion by clear distinctions; to rejoin the severed and reanalyse the commingled'.[96]

Lefebvre's social theory about space concerns many subjects that have been debated in later research, but important in the context of this book is that his trialectic model, as architecture theorist Christian Schmid writes can be used as a 'conceptual framework' or analytical instrument that can be gainfully used for a variety of questions.[97] It is important to notice that Lefebvre sees the social processes in which space is produced as a place of contradictions and conflicts. To be able to study industrial open spaces, we need a refined understanding: 'The true theoretical problem, however, is to relate these spheres to one another and to uncover the mediation between them.'[98] Lefebvre's own research included empirical studies of urban and spatial situations, and this has later inspired others – including myself – to use this lens to study empirical situations.[99]

93 Lefebvre, *The Production of Space*.

94 Stanek, *Henri Lefebvre on Space*, ix.

95 I am here leaning on a remark by Cresswell, *Place*, 38.

96 Lefebvre, *The Production of Space*.

97 The original quotation in German is 'Mit diesem Konzeptionellen Gerüst [...] steht ein analytisches Instrument zur Verfügung, das sich für die unterschidlischste Fragestellungen gewinnbringend einsetzen lässt' (Schmid, *Stadt, Raum und Gesellschaft*, 21). See also Schmid, 'Henri Lefebvre's Theory of the Production of Space'. In Goonewardena, Kipfer, Milgrom, Schmid, eds. *Space, Difference, Everyday Life*.

98 Lefebvre, *The Production of Space*.

99 Lefebvre, *The Urban Revolution*.

Authors of landscape

A long-standing idea in landscape research is that a collective of people can carry a certain culture that leads to a regional or national physical character.[100] Yet, in industrial sites we have various groups of workers and managers, as well as insiders and outsiders to the site, who act and perceive landscapes according to different agendas and mindsets. The idea of 'locals' as a group must therefore be replaced by a more refined understanding of how people participate in landscape formation. In this sense, landscape scholar J.B. Jackson's detailed studies of how people shape landscapes is useful: 'A landscape, like a language, is the field of perpetual conflict and compromise between what is established by authority and what the vernacular insists upon preferring.'.[101] Jackson acknowledges that 'ordinary landscapes' have an order that can be studied. Industrial sites do not only expose activities by engineers, managers and planners, but also acts by the people who have used and interpreted these sites in the everyday.

This book will learn from Jackson's insistence on studying heterogeneous actors in landscape in two ways. First, Carlsberg's open spaces will not only be studied as the result of acts by designers, as often in architecture history, or engineers, entrepreneurs and other industry-related actors, as often in industrial history, but by combining these. This continues an existing tradition to read Carlsberg's architectural history in relation to industrial processes, yet with much more focus on open spaces and on the youngest phases of history than previous research.[102] Second, I will include studies of how Carlsberg's open spaces were used and understood by employees and other users in greater detail than has been done before.[103] Jackson defines the word vernacular as traditions and customs that are 'entirely remote

100 Hauser and Kramleithner, *Ästhetik der Agglomeration*, 158-166.

101 Jackson, *Discovering the Vernacular Landscape*, 148.

102 This has to some degree been done by others who have studied different phases of Carlsberg's history, particularly Jørgensen. See Jørgensen, 'Tekno-økonomiske paradigmer og industrimiljøer i Danmark 1770-1970'. Jørgensen, 'Carlsbergs bebyggelseshistorie,' and Zanker von Meyer, *Die Bauten von J.C. und Carl Jacobsen;* Nielsen, *Gamle Carlsberg 1847-1897;* Michaelsen, 'Carlsbergbryggeriernes Bygningshistorie.' An early landscape-oriented study of Carlsberg is Hauxner, 'Carlsbergs landskab: et stykke kulturarv og kulturmiljø.'

103 The following draws knowledge from Kruse, *Bryggeriarbejdere i København – de første årtier* and from a web-project called 'Humlen ved Carlsberg' of the Arbejdermuseet (Workers' Museum) in Copenhagen, which presented the everyday in the breweries from the perspective of the workers, yet without focusing explicitly on open space as I will do here. The project is accessible under http://humlenvedcarlsberg.dk/

from the larger world of politics and law'[104] – a view that seems too black and white to match the reality of industrial landscapes. The following landscape biography of Carlsberg is an opportunity to study the spatial implications of different actors in closer detail, knowing that vernacular practices are not necessarily better or less strategic than those by experts and dominant actors.

Investigatory techniques and source material

To study industrial landscapes and their open spaces as dynamic and relational, I needed to combine multiple methods of study and sources of very different character. This is not because I think that this will ever provide me with a total picture, but because only by combining different types of knowledge can I begin discovering aspects of the interplay that produces Carlsberg's open spaces. The research process has moved back and forth between two types of analysis: a critical study of the contemporary discourse of Carlsberg's redevelopment and – in response to that analysis – a historical study that produced what I call new chapters in the biography of Carlsberg. The diachronic studies seek to understand the interplay between discourse, materiality and lived spaces and to unfold the mediation between these spheres.

Studies of the present redevelopment projects

To examine the contemporary urban redevelopment I have been an observer in meetings and workshops organized by Carlsberg Ltd and the Copenhagen Municipality between 2007 and 2009, and participated in public meetings and guided tours that were organized as part of the urban project.[105] I have also analysed texts by architects, planners, heritage managers, critics and others who have been involved with the redevelopment of the site between 2006 and 2009. I studied documents, meeting minutes, the competition brief, heritage reports, newspaper articles, press releases and other written material that shed light on the redevelopment process.

Next to these written documents, I consider design projects and plans as cultural utterances and treat them as surveys of Carlsberg. This involved a close study of the proposals that had been submitted to the international

104 Jackson, *Discovering the Vernacular Landscape*, 149.
105 My research began with a PhD scholarship at the University of Copenhagen in 2007.

ideas competition 'Carlsberg Vores By' in 2006-2007, where I began by screening all projects and then chose those that most clearly reflect on open spaces. From there I have studied the evolving urban plan in its different stages as well as design projects that have been made for special areas after the competition. I use both built and unbuilt design projects – the last ones then only studied through their mediation in images, plans, diagrams, models, and texts.

Design projects play three different roles in the research. First, they will be scrutinized as cultural products that reflect different interpretations of Carlsberg's landscape. Second, design projects will be used as of knowledge into the characteristics of Carlsberg. Designers often apply investigatory strategies that are not explicitly articulated to outsiders, and do not necessarily openly reflect upon them systematically. Nevertheless, design projects are investigations of a specific spatial situation. The strength of the architectural site survey, Solà-Morales argues, is that it produces a special kind of knowledge that can point to qualities that may not be traceable by other methods of site investigation. Solà-Morales calls this 'a literary knowledge' that can serve as a counterpart to other, 'causal simplifications, explanatory models and diagrammatic representations' of a site, or even replace such views.[106] While design is not the only field that produces literary knowledge, one of the special things about it is to communicate knowledge by visual and spatial representations rather than just text. Such representations bring in aspects that cannot be grasped by text and therefore potentially add dimensions that would otherwise remain unseen.

The third role that design projects play in this book is as historical sources into Carlsberg's past. Previous design intervention affects the present materiality of this landscape, and if we know more about design projects of the past, we can better understand the open spaces in the present. Even design projects that were never realized can be sources of knowledge about a landscape. Such projects can become part of the cultural imaginary connected with the particular location, and thereby steering the spectrum of possibilities for future intervention. Including design projects as sources in a landscape biography may be a key to overcome a difficulty that has been noted in previous research: While aiming at impacting practice, it has sometimes proved difficult to establish a dialogue with designers in projects where landscape biographical studies have been made available to practice.[107] By emphasizing open

106 Solà-Morales, 'The Culture of Description,' 19.
107 Roymans et al., 'Landscape Biography as Research Strategy,' 352-353.

space, discussing and including design proposals as sources, the dialogue between designers and other actors in urban redevelopment as well as researchers can be strengthened.

Studies of the production of open spaces

The diachronic investigations of Carlsberg combine a broad spectrum of source material to embrace how space is produced in the relationship between materiality, practices and discourse – the Lefebvrian triad. I have combined primary and secondary, dominant and non-dominant written, material, visual and oral historical records. The diachronic analysis relies on many sources: archival photos, drawings, maps, newspaper reports, design proposals and previous historical research. Most of the historical records were found in the Carlsberg Ltd archives. Further, I used historical maps from Generalkonsulen, Kort- og Matrikelstyrelsen, Hærens Arkiv, Det Kongelige Biblioteks Fotosamling (the photograph collection in the Royal Library in Copenhagen), Frederiksbergs Stadsarkiv and Sognekort Hvidovre, Frederiksberg, Copenhagen.

I studied the morphological components and structures of Carlsberg's open spaces through on-site observation, photographs and analytical drawings, as well as studies of maps and aerial photos, both historical and new. I combine this structural perspective with an examination of how the space evolves as it is perceived from within. Readers will be invited to a walk I did in 2008, before I knew the site well. I used no map, but – inspired by what urbanist Guy Debord named *dérive* – took advantage of my curiosity as an outsider and photographed and wrote about my first impressions and sensations.[108] This first bodily encounter came to colour my perception of Carlsberg later on.

To learn about the everyday in the brewery, I relied on a broad range of published personal histories, which are quoted in the text when used. The brewery company allowed me access to pass through the strictly gated entrances that were there when the research began, and I moved freely as on the restricted brewery site in the open air from July 2007 until December 2008, when most of these gates were opened to the public and I could enter without special permission. These visits enabled many informal talks with employees I came across, who generously invited me in to different buildings and facilities and explained their everyday in the breweries, drew maps of the open spaces, their colloquial names and meanings, together with me,

108 Debord, 'Theory of the Dérive.'

which I later confirmed with Ulla Nymand, director of Carlsberg's archive.[109] The employees also shared old Carlsberg myths and stories with me, some of which I will now pass on to the readers of the following chapters.

109 Nymand, personal communication, 7 July 2010.

2 Site

Just like many other production facilities, the Carlsberg breweries have been constructed as a distinct site, separated from its surroundings by fences and gates. For functional reasons and to follow the tax laws for beverage production the area was restricted and only people who had clear appointments could enter, right up until most production was relocated in 2008. Yet, while having such a closed character, the breweries also evolved as part of the city of Copenhagen and of larger networks that distributed knowledge, raw materials, goods, money and other industrial assets. This paradox – being at once separate and part of a larger urban landscape – is characteristic for former industrial sites and constitute a planning challenge. Further, this Janus-faced character shows that a site is not an entity that merely *is*. Concepts like industrial sites, heritage sites, urban development sites, and competition sites are frequently used in daily practice, often without further discussion. Yet, as this chapter will discuss, the very act of defining the boundary of a site is a matter of interpretation and taking positions – be it in prospect and in retrospect.

The plural form of the company name, the Carlsberg Breweries, already testifies that this may not be one site, but many.[1] The official name of the urban project that begun in 2006, however, suggests something else: Carlsberg City. The name seems to try to shake off the fact that the former brewery is engulfed in the surrounding city to all sides and dependent on it in many ways. This chapter explores the tensions that characterized the way that Carlsberg was defined as a site during the redevelopment project. To contribute in this discussion, I propose new narratives about Carlsberg using a relational concept of site.

Carlsberg – Becoming a building site

The company, now known as Carlsberg Ltd, is today one of the world's largest breweries. It has played a significant role in Danish cultural history, having formed charitable foundations that have contributed to Danish academia and arts for 150 years.[2] The brewery's success has enabled it to expand to

1 Carlsberg got its plural name after the merge of Old Carlsberg and New Carlsberg in 1907.

2 Among the numerous foundations that have been established in close connection to the Carlsberg Breweries or companies that are now part of the Carlsberg Group, the largest and

approximately five times the size of the original premises on Valby Hill, first built in 1846. During the last century the Carlsberg property has become engulfed by the growing city.

Image 2.1 The Carlsberg property is surrounded to the south by a railway ditch; to the east by a district of single-family housing dating to the late 1800s; to the north-west by Søndermarken, which is part of the gardens of Frederiksberg Palace and single-family housing dating from the 1700s to the 1900s; and to the east by five-storey housing blocks in the dense neighbourhood of Vesterbro, built in the early twentieth century.

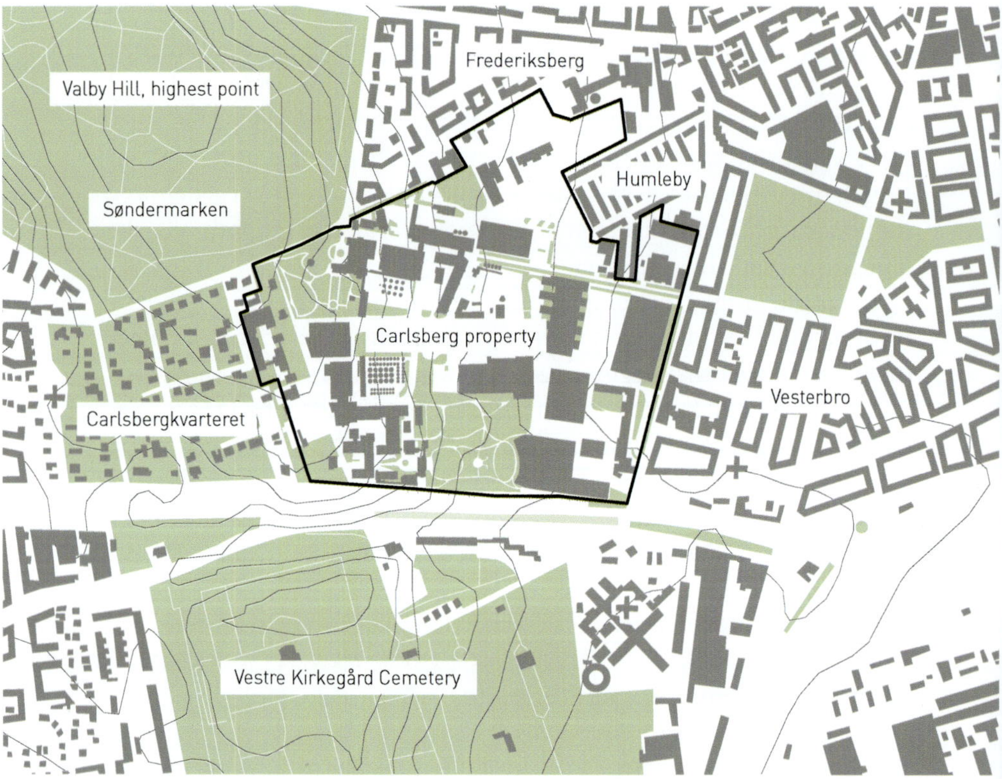

best-known are the New Carlsberg Foundation and the Carlsberg Foundation. See Glamann, *The Carlsberg Foundation*; Glamann, *Øl og Marmor*; and Glamann, *Bryggeren*.

Carlsberg's industrial landscape has been continuously rearranged according to the changing production and distribution methods in brewing. In the last decades the role Carlsberg's Copenhagen premises has diminished in proportion with its worldwide activities. The founding premises in the Danish capital supplied the home market, yet in 2006, the company decided to move the Danish production to a newer brewery in the Danish town of Fredericia. Carlsberg's global headquarters remain on the iconic Copenhagen site, as part of the new urban district. While relocating Carlsberg decided to commission the overall urban planning; their goal was originally to sell properties once new buildings and urban space design projects had been realized. The company is known for selling beer, but not for being a developer. Yet Carlsberg decided to play such a strong role in the urban project because they believed that doing so could minimize potential failure, criticism and damage to the corporate brand.[3] The relocation to Fredericia entailed the risk of bad press; not only were many workers dismissed, but the massive building plans would significantly change this popular historical site. Carlsberg's image on the Danish market as a carrier of cultural values could be threatened, the company knew. To prevent this from happening, the urban redevelopment project included thorough analyses, a close collaboration with the municipality and heritage authorities and not the least intense public communication initiatives.

During the 1990s Carlsberg Ltd had been a growing company with production in Denmark and elsewhere, but it was not possible to physically expand the Copenhagen property because of the surrounding city. Moreover, the site was full of old built structures and it was difficult to find room for the most recent technological inventions and most rational distribution methods, which however happened between 1980 and 1995 Carlsberg Ltd. In these last two phases of reorganization of the Carlsberg breweries several buildings were demolished to make room for new methods for brewing and distribution.[4] This can, for example, be studied by comparing aerial photos from 1954 and 1995 (2.2-2.3). During these four decades, the Carlsberg site changed from having many small units with small buildings; to increasingly consisting of large open spaces and expansive buildings.

3 Ulvemann Explorative and Carlsberg Ltd Properties. 'En kvalitativ undersøgelse vedr. image af Carlsberg og Apollo.' This decision was also informed by a previous project – the redevelopment of the Tuborg brewery site at the harbour of Hellerup north of Copenhagen. This urban redevelopment project has been heavily criticized in public as an urban failure.

4 This had been approved by Copenhagen Municipality in 'Lokalplan 44.'

Image 2.2 Aerial photograph of Carlsberg, 1954. Throughout the twentieth century the Carlsberg site has been repeatedly reorganized due to changing production rationalities. The black squares highlight places where two of these changes take place.

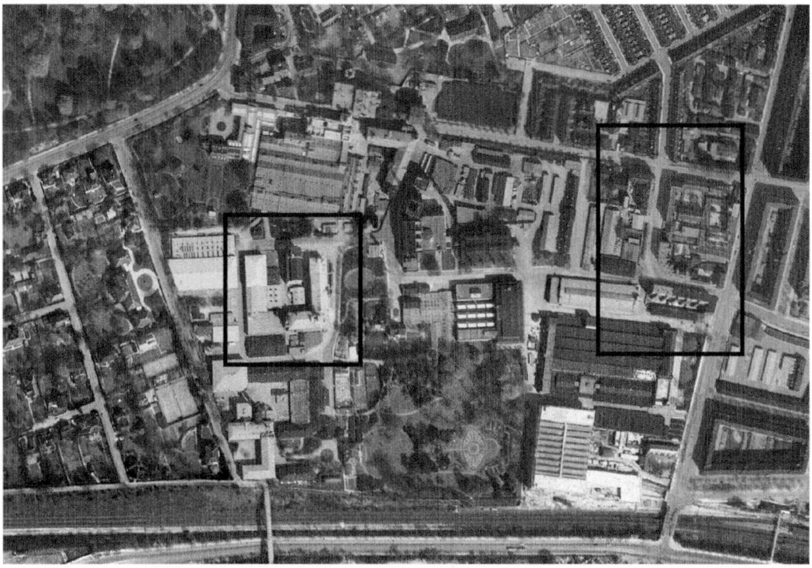

An integrated part of these demolition activities was reusing old materials and equipment, for instance, when new technology replaced fermentation casks in the 1980s, the granite from these casks was recycled as paving casks.[5] These practices were never talked about as heritage, but rather merely as a way of circulating materials for practical reasons. However, it was known to many of the workers that I met on-site in 2008 that the paving came from the casks and thus it became part of the oral history among insiders and a way for them to connect the present with the past.

In 2006, when the urban redevelopment was announced, the situation was radically different; Carlsberg Ltd did not act as an industrial firm but rather as a developer with economic interests to sell and brand Carlsberg City as distinct from other emerging urban districts on the market. That rationale is reflected in the brief for the ideas competition, which grapples with situating the district in the city: 'Entrants should work with the Carlsberg district both as a part of Copenhagen and as a separate new city

5 Nymand, personal communication, 31 October 2008.

Image 2.3 Aerial photograph of Carlsberg, 1994. Carlsberg during the last major reorganization phase as a brewery. In the areas marked with black squares, nineteenth-century buildings had been demolished in order to make room for new means of distribution.

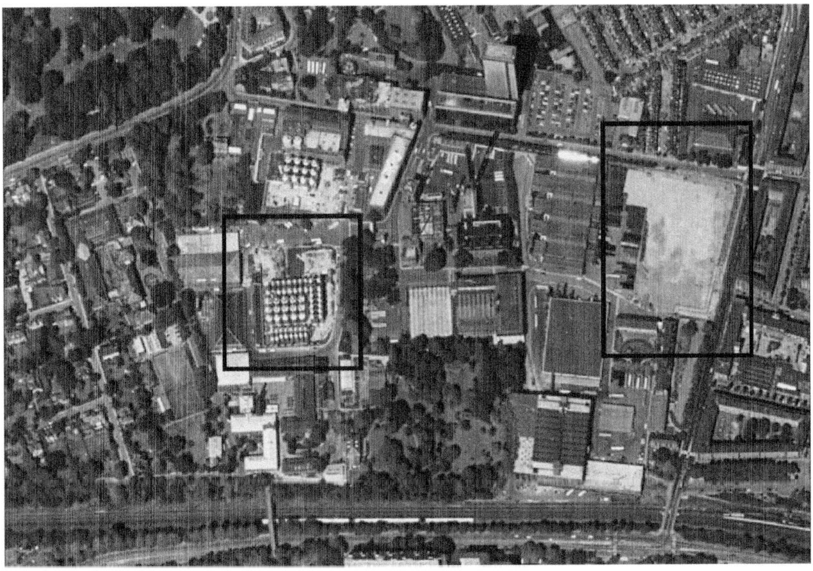

district. Thus spatial relationships will be crucial, not only for the area's integration with the city around it but also for the Carlsberg district's own identity.'[6]

The brief paid much attention to Carlsberg's adjacent areas and highlighted the challenge of integrating a formerly contained production plant into the city.[7] On the other hand, Carlsberg Properties Ltd consequently described the site as a city; in the beginning the slogan 'Our City' was used, referring to the company's commercials in the Danish market for 'our beer'.[8] The stated aim of Carlsberg Ltd was to create an urban district with a 'clear identity'.[9] Several design teams in the ideas competition worked to comply with this need for distinction by imposing a homogeneous urban

6 Carlsberg Ltd Properties, 'Vores By Carlsberg,' 43.

7 Ibid., 42

8 The official name for the urban project is 'Our City' (which refers to the company's Danish beer slogan 'Our Beer') and, since 2009, 'The Carlsberg City.'

9 Carlsberg Ltd Properties, 'Vores By Carlsberg,' 41.

fabric on the entire site, clearly distinguished to the surrounding districts. The winning proposal, for instance, installs the same urban form principle for the entire Carlsberg site: three- to five-storey city blocks that create narrow streets and trapezoid city squares, each combined with a high rise that serves as a landmark when seen from afar (0.3).

To transform the brewery into one unified site with one built form principle appears as a radical decision when we look at the Carlsberg site as photographed in 2005. When production was relocated, the brewery was a conglomeration of different areas, spatially organized according to different geometries and built in different sizes and with different relationships

Image 2.4 The Carlsberg brewery site as photographed in 2005, just before the urban project began. The picture shows a conglomeration of different areas spatially organized according to different geometries; built in different sizes and with different relationships between buildings and open space.

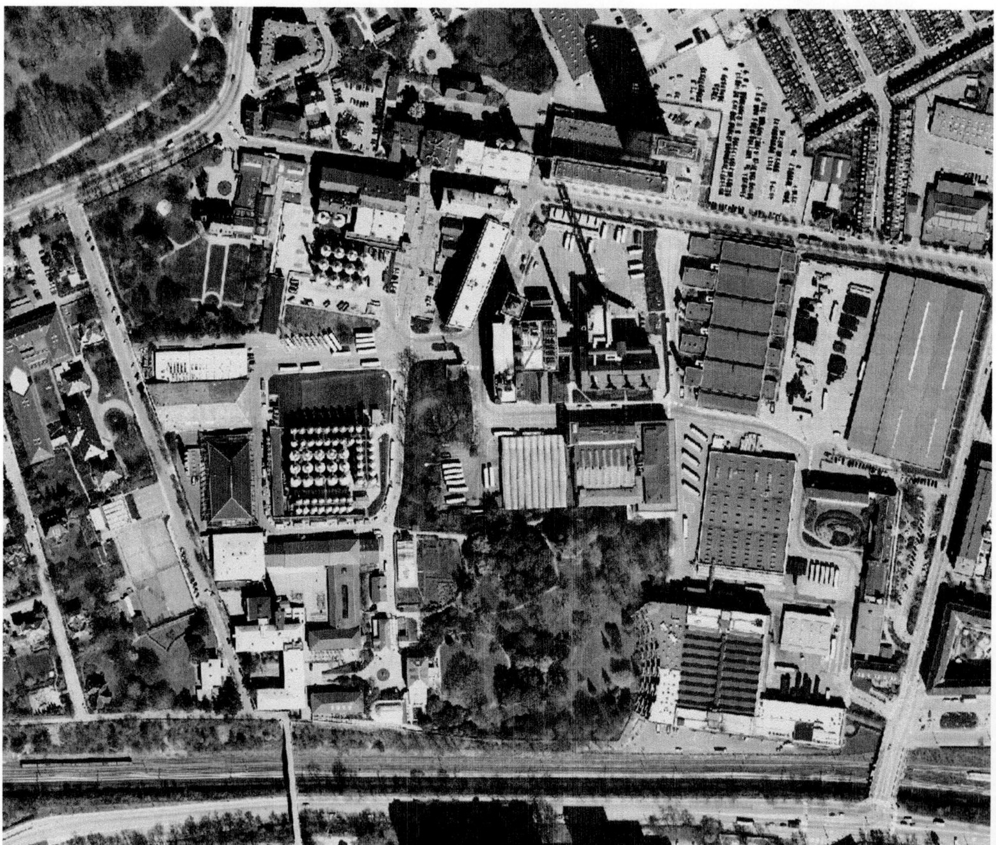

between buildings and open space (2.4). The eastern, topographically lowest parts of the brewery are characterized by large asphalt surfaces and massive buildings, while the higher, western and northern parts have varied combinations of buildings with different shapes and sizes and two large gardens.

The different physical structure that appears from buildings and terrain modelling separates the site into a number of seemingly distinct areas. Some of these areas have official names that arise in the official maps and literature about Carlsberg – the MB Brewery, Old Carlsberg and New Carlsberg. Other areas, namely those that are built in the twentieth century, are less known to the public and are often presented without being named.[10] This structure with several areas that do or do not seem to adjust to each other is also part of the experience that first met me as a researcher.

A walk

I would like to invite you to a walk through the Carlsberg site in August of 2008, when I was more curious than informed. I let myself drift through the area, directed by instinct rather than a predetermined route, thrilled to explore this unknown terrain in my city. By then, the urban redevelopment process had just begun. Some production activities had already been relocated, but beer was still being fermented and bottled here. Carlsberg still had gated entrances that ensured that no beverages were taken out of this separate tax zone. Although only few were permitted entrance, the Carlsberg brewery site was known to most Copenhageners, who may have been to the Visitors' Centre when they went to school. The two main roads, Ny Carlsbergvej and Gammel Carlsbergvej, have always been public and allow people to relate to the area. What was behind the fence was less known. And now I was permitted entry.

I went in through Gate C, which did not look like any main entrance, although being one. I was told to get permission to enter at the guard building and felt terribly alien right away. As I walked in, I instinctively followed a broad road that led me through curves and vistas. Yet, I constantly had to step back, every four minutes or so, to make room for one of the large tankers that made a lot of noise and took up all the space of the broad road. This walk went to spaces that were not immediately comprehensible to me as an outsider; In one moment I walked on fragments of a typical

10 This book will use official names stemming from the literature as well as colloquial names used by brewery employees.

Image 2.5 First impressions of the Carlsberg breweries, pictures taken by the author during summer 2008.

Copenhagen pavement and in another I spotted a symmetrically arranged garden with a monument. My walk was sometimes roofed as if someone wanted to protect me from the hot sun. At least they sheltered me from the noisy trucks that kept pushing me out of the *derivée* I wanted to take. There were also a lot of odd spaces that caught my attention – such as the one with strange thermoplastic patterns on the streets that communicated a message I was unable to follow.

There was a gate that made no immediate sense because it was just standing there, detached from everything else and leading to nowhere particular. There were mazes made of crates. I felt addressed all the time by glorious Latin inscriptions, sculptures that stared at me, golden ornamentation and just the very largeness of many of the old buildings.

The walk took me through many different atmospheres: a quiet garden, places with a strong smell of hops or spices, and wide open spaces in which my eyes could rest. The first part of the route was on flat ground, while the terrain sloped as I went along. The open spaces seldom seemed arranged for experiencing from a single standpoint, though; perception in motion seemed to be encouraged. On the other hand, some places were difficult to navigate by foot and much better suited for the massive vehicles that drove through again and again. Over and over people addressed me to wish me welcome, as if I had entered someone's home. They proudly told about their work and history in the company, sometimes dating generations back.

This walk left me with the impression of a conglomerate of different open spaces. A route did connect these open spaces but their functions seemed incomprehensible. The kinaesthetic experience of walking up the hill was striking, and as I walked, I passed many terraces, slopes and retaining walls shaping and opening up spaces and the passages between them. Utterly different open spaces appeared to juxtapose in unusual ways – sometimes hinted at through focal points, where one space floated into another, at other times disrupted. This spatial complexity seems to have been the starting point of several alternative readings of the Carlsberg site, which, unlike the winning proposal, celebrated the conglomerate instead of trying to create unity.

Heterogeneity in retrospect and prospect

A competition entry that was neither awarded nor much discussed in the public was made by British Studio Egret West (2.6). Their proposal 'Collage Village' sees Carlsberg as the mix of many areas with distinct morphological characteristics, in terms of buildings and open spaces.[11] Each of these areas, Egret West argues, emerged in different formation processes and should be cultivated on its own premise in the future, too. The areas should be developed for different uses and users in different tempi and follow different forms of ownership.

Collage Village strives to enforce the heterogeneous character of Carlsberg, which is seen as an opportunity for varied urban experiences in the future:

> The challenge at Carlsberg is how best to embrace this eclecticism and design for the multifaceted city. In answering this question, Carlsberg should stand out as a different kind of urban model. One that embraces imperfection. One that is accepting. Not sanitizing. One that allows potentially conflicting patterns or users to sit next to one another. One that avoids a 'grand vision' or a single-handed master plan or a unified architectural language.[12]

11 This title is a reference to Colin Rowe and Fred Koetter's book *Collage City*, which reappraised heterogeneous urban form.

12 Studio Egret West, 'Collage Village,' 2.

Image 2.6 Studio Egret West, who designed this competition proposal called Collage Village, conceptualized the existing Carlsberg site as disintegrated into many separate areas, each with its own characteristics, both historically and in the proposed future. According to the designers, these areas should not develop according to a comprehensive masterplan. Rather, each area in the "collage village" should evolve slowly in a process steered by different users, owners and actors, who would develop different programs, spatial characteristics and different building types for the area. This, the designers proposed, could reflect and stimulate a multifaceted city.

In terms of demolition, reuse and preservation, Studio Egret West proposed to 'keep everything (or at least to start with)'[13] and emphasized Carlsberg's morphological diversity. Notably, they propose that this spatial diversity is a way to enforce differentiated aesthetic experiences as well as potentially to reflect and even fuel social and cultural diversity.

Relational site in prospect and retrospect

Another question that exposed different views in the redevelopment process concerns defining the extent of Carlsberg. While the urban redevelopment project was still in planning, in 2008, the Heritage Agency of Denmark surveyed Carlsberg's buildings and gardens to select which ones to be listed. This survey assessed exclusively that which was inside of the property line of Carlsberg Ltd in the present, although the brewery has been much larger in the past. When asked why it did this, a representative from the Danish Heritage Agency answered, 'Well, right now, it is Carlsberg that we are dealing with here.'[14] The heritage experts saw Carlsberg as a stable unit whose size required little consideration. This perception fitted perfectly with the corporate aim of creating a direct link between the – glorious – past of the company and its future development into a profitable urban district, as expressed in the brief: 'In the development and conversion of the site, the narrative of the history of Carlsberg must be continued.'[15]

This fixed perception of the historical location of Carlsberg was broached in another heritage designation by the Heritage Agency of Denmark, emphasizing the interpretative nature of the concept site.[16] In 2007 Carlsberg was selected as the first of 25 'Industrial Sites of National Significance' – a new category in Danish heritage management that was taken as an opportunity to rethink how to delimit the breweries.[17] The authors of a preparatory report that guided this designation highlighted the importance of expanding the interest from industrial complexes in themselves to study their relationships with other places. The report recommends looking at the 'physical structures that interconnect particular

13 Ibid., 3

14 Brorsen, personal communication.

15 Carlsberg Ltd Properties, 'Vores By Carlsberg,' 28.

16 Jørgensen, *Carlsberg som nationalt industriminde.*

17 Heritage Agency of Denmark. 'Carlsberg som nationalt industriminde.' Communication to Carlsberg Ltd Properties, 22 September 2006.

industrial complexes in industrial quarters',[18] such as roads, harbours, wires, pipes and railway lines:

> The industrial quarter, seen as an entity, consists of numerous structures and relations, and embeds the particular industrial environment in a physical and contextual setting. These structures and relations – physical and in terms of content – are thus preconditions and condition for the emergence, existence and resilience for each industry.[19]

Interestingly, when Carlsberg was selected as an industrial site of national significance, it was defined as a site much larger than the contempoary property line. 'The Carlsberg breweries' now also included two adjacent housing areas that have not been built as part of the breweries, but have housed some of its employees,[20] a nearby church built by brewer Carl Jacobsen in 1891[21] and the brewery to the north of Carlsberg's property (Kongens Bryghus), which only belonged to Carlsberg Ltd for a short period around 1990.[22] Notably also, a section of the adjacent railway is part of this heritage site definition; it was important to Carlsberg's transportation from the 1930s to the 1970s. The boundary of Carlsberg as it was defined as an industrial site of national significance appears to be rooted in a historical analysis that considers infrastructural networks and the social and cultural life of buildings after their realization.[23]

18 Selmer, *Industrisamfundets Kulturarv*, 8. Translated by Marion Frandsen. Original quotation: 'fysiske strukturer, der forbinder de enkelte industrikomplekser med hinanden i større industrikvarterer'.

19 Ibid., 11. Translated by Marion Frandsen. Original quotation: 'Industrikvarteret betragtet som en helhed, bestående af en række strukturer og relationer, placerer de enkelte industrimijøer i en fysisk og kontekstuel sammenhæng. Disse strukturer og relationer – fysiske som indholdsmæssige – er således forudsætninger og vilkår for de enkelte industriers opståen, eksistens og levedygtighed.'

20 While neither of these housing districts were built for or by Carlsberg, employees of Carlsberg have lived in some of these houses. Humleby was built by a worker's association. See Bülow Jensen, 'Humlebyen Arbejdernes Byggeforenings.'

21 J.C. and Carl Jacobsen, their wives and children are buried in this church.

22 Nymand, personal communication, 7 July 2010.

23 This expanded site definition, however, appears to have been met with some resistance. When the designation 'Industrial Sites of National Significance' got its most concrete manifestation – a book with a foreword by the director of the Heritage Agency – the expanded site definition of Carlsberg had shrunk significantly. This probably demonstrates the effect of habitual thinking; the usual perception of Carlsberg as the property line is so common that its reintroduction likely entered the book without anyone noticing that it did not correspond with the first report. See Jørgensen (ed.) *Industri Industri*, 49.

Image 2.7 The 'Time Will Tell' proposal for the international ideas competition 'Carlsberg Vores By' concentrates its effort on vitalizing the borders by programmes that relate it to the surrounding city. To the west: windmill parks that serve Carlsberg and other districts. To the south: meeting places, a new path and a fun fair. To the north: housing with workshops and small shops for people to visit. To the east: parking and a drive-in cinema. The public functions of these borders are thus thought to inscribe Carlsberg within the city in new ways.

This heritage surveys was made after the international ideas competition 'Carlsberg Vores By'. Yet nevertheless, the competition entries express similar questioning of Carlsberg's borders and an emerging interest in how Carlsberg participates in networks and processes beyond the property line. Several design teams proposed ways to embed this formerly restricted site more strongly in the city and some broke out of the brief's limit of requested intervention, most radically a group of Spanish students, who won second prize in the competition. Their proposal 'Time Will Tell,' Alberto Álvarez Agea, Ana Zazo Moratalla, Manuel Álvarez-Monteserín Lahoz, Ana Peñalba Estébanez and María Mallo Zurdo avoided defining a program, form and final goal for the entire site (2.7). As the title indicates, the proposal prescribed a slow change without a final goal and with hardly any design intervention on the actual property of Carlsberg Ltd. Rather, to help a process of appropriation by new users along, the young architects suggested that what should be developed first is precisely the borders separating the former brewery from what is outside it. The designers introduce a windmill park to inscribe the Carlsberg site in a new network of renewable energy production and use. Also, new social meeting places, new walking paths a spectacular new drive-in cinema and funfair are thought to inscribe Carlsberg within the city in new ways. These ideas are not supported by a thorough analysis of the local conditions and do not bear signs of visits to this particular site, but should rather be seen as conceptual comment to the brief of making a coherent plan for new buildings to fill the entire site.

Coming from different fields and seemingly unaware of each other, both designers and heritage experts questioned the fixed definition of the Carlsberg site. Rather than one site, they seem to argue, Carlsberg can be seen as multiple sites, engangeled both inwards and outwards, depending on the point of view. This interpretative act characterizes both the different heritage surveys and the designerly investigations and prospects.

Their different takes on how to define Carlsberg as a site underlines that Carlsberg, like any site in the city – and industrial sites in particular – is not an essential thing out there, but rather a 'relational construct'. This notion has been introduced by urban design thinkers Carol Burns and Andrea Kahn to describe how urban sites are not stable, but 'constantly reconstituting as part of specific contexts in which they must be understood'.[24] The heritage professionals and designers discussed here take an active part in defining contexts for Carlsberg some by voicing alternatives to an essential site definition and thus opening up for alternative future scenarios.

24 Burns and Kahn, *Site Matters*, x.

Chapters in the biography of Carlsberg as a relational site

The surveys that emphasized Carlsberg's spatial heterogeneity and entanglement in a larger urban landscape discussed above have inspired, the following new chapters in Carlsberg's biography. How can Carlsberg be understood as a site evolving in dynamic relationships with the local and even global contexts it is part of? To understand this, I present certain histories, each focussing on one aspect of Carlsberg's becoming as a site and different contexts in which to view it. These histories apply a longue durée-perspective and thus seek to transgress the temporal categories that are often connected to industrial sites. To think of anything as the 'industrial period' is a productive analytical construct but of course not a stable container of a separate past. The risk of specialized examination of one period is that it can restrain us from discovering aspects that are not related to industrial rationales and that are older than the industrial period.

Carlsberg in the story of a hill

Copenhagen's topography is flat, with many bodies of water and only a few hills. The only slope that can be seen from the city core is Valby Hill; on which the Carlsberg site is located. Valby Hill is a sandy, Ice Age formation with subterranean water sources with chalk-rich aquifers. Its slopes, protected from the windy shore, form a slight microclimate of milder temperatures than the city below. A geological fault stretches beneath Valby Hill and bears the company name: the Carlsberg fault. The Carlsberg brewery was first built Valby Hill in 1847. Prior to founding this site, the ambitious entrepreneur J.C. Jacobsen was one of approximately 50 brewers in the dense city of Copenhagen.[25] He decided to relocate his brewery to an area where the water supply was better than in the dense city. This is the main reason that Carlsberg was founded outside of the city ramparts, as the first willing to pay the extra taxes required to bring the goods back into the city for sale.[26] Jacobsen named the new brewery after the hill (berg) and his then five-year-old son Carl Jacobsen – *Carlsberg*. While both the official communication of the contemporary urban redevelopment project and the seminal historiography present J.C. Jacobsen as a pioneering, creative

25 The most exact number is from 1833, when there were 46 breweries in Copenhagen. Tullberg, *Dansk Bryggerifortegnelse*; Strømstad, *J.C. Jacobsens Bryggergård*.
26 Glamann, *Bryggeren*, 76.

figure[27] – which he was – he was also part of an urbanization process that happened at the time.[28] The urbanization begun with the establishment of the Frederiksberg palace gardens (2.8) and continued driven by changes in the regional road system, which can be studied on the next pages (2.9, series of images).

Image 2.8 The Frederiksberg palace gardens, as depicted on a map from 1795. The Søndermarken garden and Frederiksberg Palace to the north-east. The black line shows where the Carlsberg property is situated today, and the Farm Bakkegaarden, which was there when the map was drawn, can be seen underneath.

27 Glamann, *Bryggeren*; Glamann, *Øl og Marmor*; Zanker von Meyer, *Die Bauten von J.C. und Carl Jacobsen*.
28 The following draws much from Lund Jørgensen, *Valby Bakke*.

Image 2.9 Transportation and urbanization processes.

1697

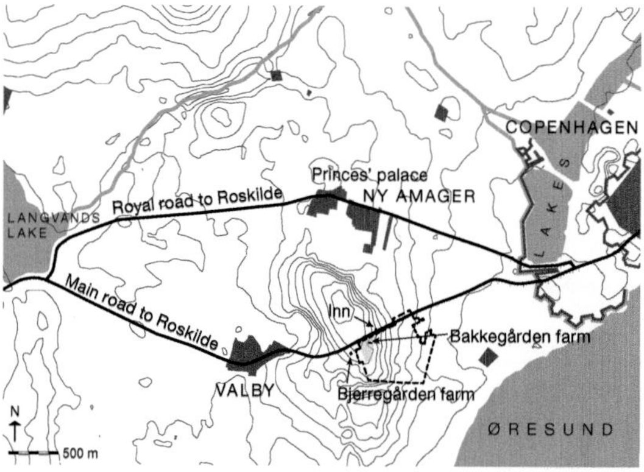

The road connecting Havn (later Copenhagen) and the other central regional trading town Roskilde was crucial for the transportation of goods and is part of the growing national royal road system. On the site that will later be Carlsberg are highly productive farms. Close to Carlsberg are inns for the many travellers en route to the gated city of Havn.

1705

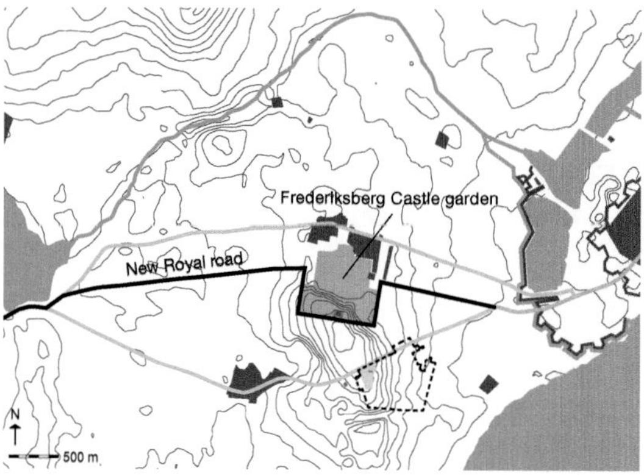

King Frederik IV builds a residence with a large garden on the top of Valby hill, and constructs a new road to service the palace, completed in 1776. The old Roskilde Road along the present Carlsberg site looses its important function and the inns start loosing their customers. The new road leads over the hill in straight lines without adjusting to the terrain.

1795

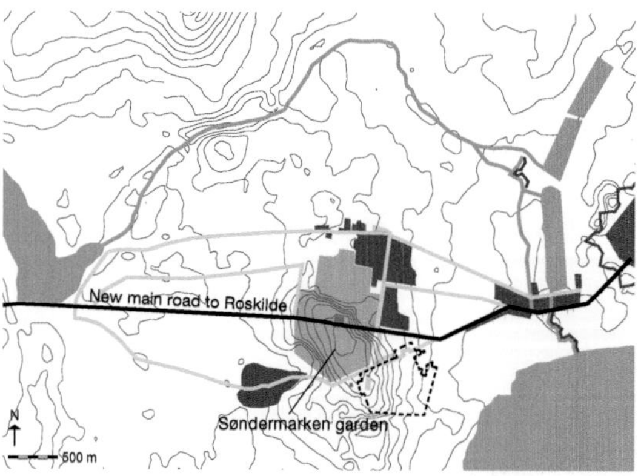

Inspired by the palace, middle-class families establish summer residences in the former inns and later build villas on this part of Valby Hill, outside of the gated city. The site of the future Carlsberg area loses infrastructural significance and new users of the land emerge.

1848

Denmark's first railway is built on the busy route between Copenhagen and Roskilde. A massive railway channel though the southern side of Valby Hill is excavated just before the first Carlsberg complex is built. Carlsberg was, however, not connected directly to the railway, but used roads and shipping for transportation.

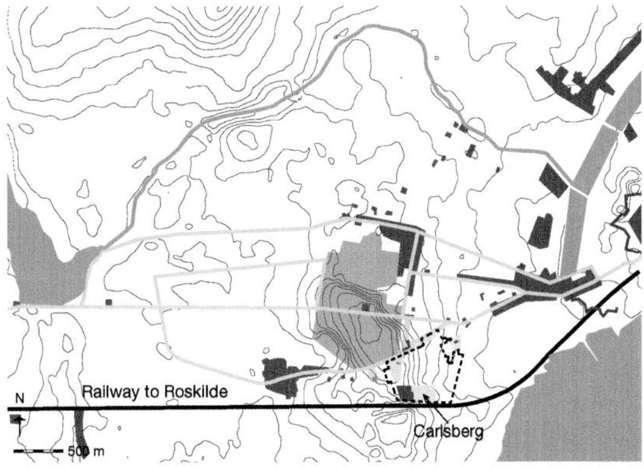

1879

Copenhagen's military demarcation line has now been redundant for two decades and the city has begun to expand massively. Due to extension and construction work of the railway the tracks are not used for train traffic (1867-1919), but as a walking path.

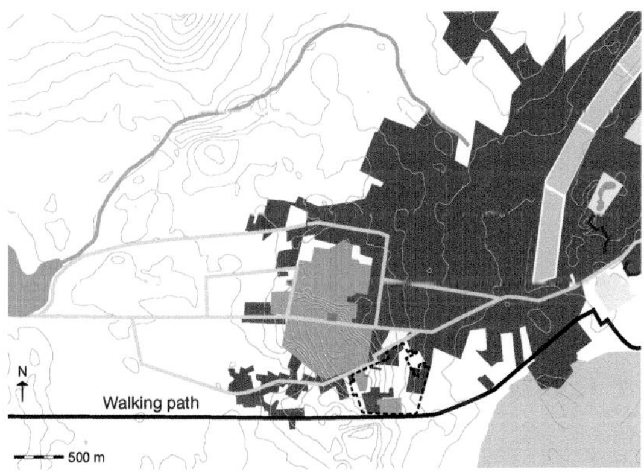

1930

The Carlsberg site becomes connected to the railway by four sidings in 1937 and later by its own station (1951) Hof and now has and has large markets abroad. The brewery is not engulfed by built-up space.

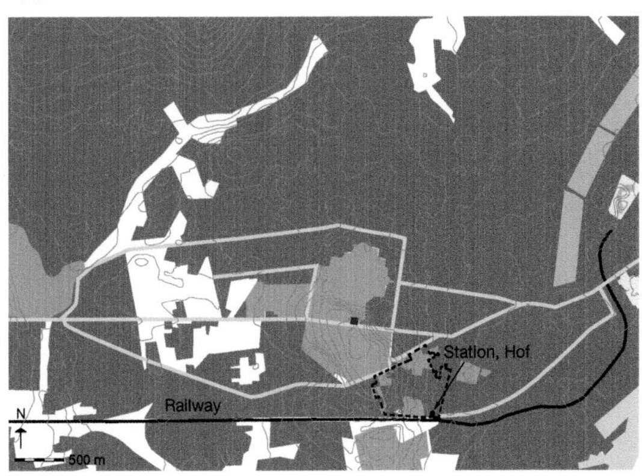

2007

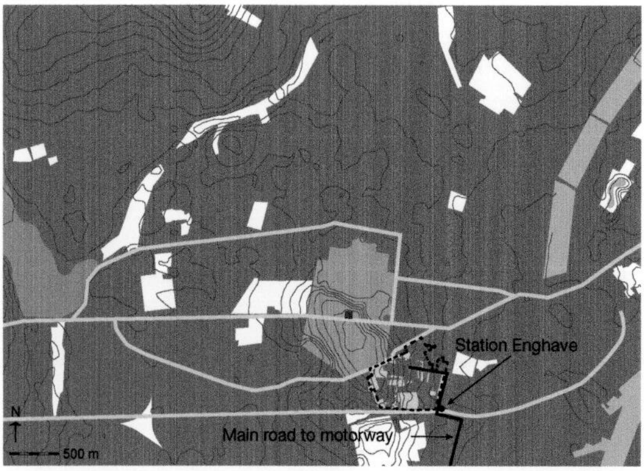

Carlsberg has stopped using the train station and instead uses trucks and tankers for transportation of goods on the motorways. In 2006, the company announced that it will redevelop the Carlsberg premises into an urban district and attempts to attract a new metro station to the area, but without success.

Situation as planned during the urban project (2009)

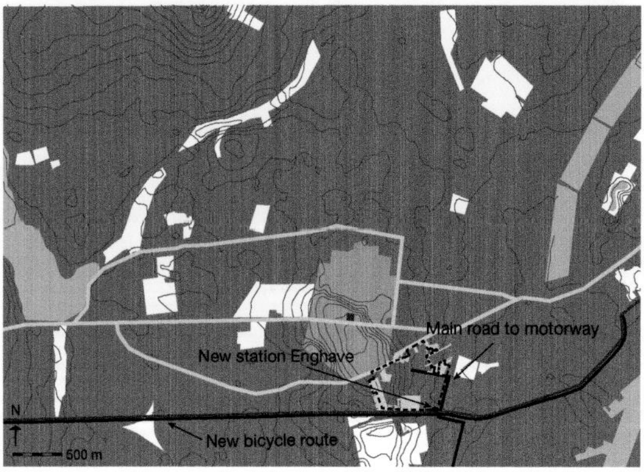

The Danish Rail agrees to move the nearby S-train-station Enghave (300 metres away) to the new city distric in Carlsberg. This station opened in 2015 and is called Carlsberg. The municipal plan also prescribes new bus routes through the new Carlsberg district, subterranean parking cellars and a new bicycle connection.

The Carlsberg site can be set with a context that started with Frederick IV's residence; Frederiksberg Palace and its gardens, dating from 1699 to 1726, are part of a cultural phenomenon that spread among the affluent all around Europe: the construction of the *villa suburbana*.[29] Like its Italian models, Frederiksberg Palace was established as a secluded residence on the highest point of Valby Hill, with a view over the distant city, and to the sea as well, before landfilling in the harbour during the nineteenth century marred the view to the sea. But quite contrary to the dream of a secluded *villa suburbana*, the palace initiated an urbanization process as a response to the growing perception by Copenhageners of their city as a gated problem. Epidemics, difficulties in getting fresh water and a growing lack of space became a major topic for discussion among the city's public.[30] Those who could afford it escaped the dense congestion of Copenhagen and sought the fresh air beyond the city ramparts. For a generation of young merchants, artists and intellectuals, Valby Hill was a pleasant alternative to the city, and they celebrated it as an arcadia with 'flowing fields of corn, wide meadows and views over Kalvebod beach' (2.10).[31] The lively seasonal atmosphere in the summer residences eventually entailed permanent settlements, and step by step a town of its own, Frederiksberg, grew around the palace and older villages on Valby Hill.

An important place in the history of Valby Hill is Bakkehuset (Hill House), a former inn for travellers along the old road that connected Roskilde and Copenhagen.[32] When the Frederiksberg Palace officials changed the road system and rendered the old Roskilde Road less important (2.8, 1705),[33] the novelist Knud Lyne Rahbek moved in (in 1787) and after he married Karen Margarete (Kamma) Rahbek they lived here together. The couple, who named the house Bakkehuset, hosted dinner parties for the new cultural elite, a group that was important in the cultural exchange of what has later been called the Golden Age in Danish arts.[34]

29 A survey was conducted of industrial buildings in the Copenhagen region; see Jørgensen, *Industribygninger i Københavns Kommune*.

30 Rasmussen, *København*.

31 Nystrøm, *Frederiksbergs Historie*, vol. 1, 13. Translated by the author. Original quotation: 'bølgende kornmarker, vide enge og udsigt over Kalvebod strand'.

32 Nørballe, Clante and Fisker, *Omkring Bakkehuset*.

33 Bach and Møller-Rasmussen, *Gennem Veje og Alléer på Frederiksberg*.

34 For example, the reformist theologian Nikolai Frederik Severin Grundtvig, physician Hans Christian Ørsted and the philosopher Søren Kierkegaard. See Lund Jørgensen, *Valby Hill*, and Dreyer, 'Kamma Rahbek og Johanne Langberg.'

Image 2.10 A generation of young merchants, artists and intellectuals dreamt about and celebrated Valby Hill as an arcadia. Valby hill seen from just outside the Copenhagen city walls. The first Carlsberg complex (later renamed Old Carlsberg) to the very back left of the picture.

Like the Rahbeks, several new residents named their houses in line with positive connotations to the hill, such as Skraaningen (The Slope), as – a generation later, and in keeping with the same idea, the Carls*berg* brewery. The brewers J.C. and Carl Jacobsen cultivated close contact with artists, writers, scientists, intellectuals and politicians[35], many of whom had prepared the ground for the vivid cultural environment in the area adjacent to the new brewery.

A passionate connection to Valby Hill can also be traced back to the first people who used it: Bronze Age settlers about 3000 years ago, who were likely to have given the hill its first known name, Solbjerg (Sun Hill).[36] The name suggests that rituals for worshipping the sun may have taken place here. In his seminal book on Copenhagen's history, architect and urban

35 Among the many who have visited Carlsberg are the author Hans Christian Andersen, the actress Louise Heiberg, the sculptor Bissen, and the physician Louis Pasteur. See Glamann, *Din hengivne Jacobsen*; Zanker von Meyer, *Die Bauten von J.C. Jacobsen und Carl Jacobsen*, 18-19.
36 Nystrøm, *Frederiksbergs Historie*, vol. 1; Lund Jørgensen, *Valby Bakke.*

historian Steen Eiler Rasmussen argues that Valby Hill had a particular significance to early settlements in the region due to having an 'even slope and a long contour line, like a large ridge'.[37] Four or more burial mounds have been located on the highest point of Valby Hill in the Bronze Age. These remained visible in the landscape for centuries like three-quarters of all the known burial mounds in Denmark, no obvious traces are visible above ground today.[38] Most of these mounds were erased when the absolutist king built Frederiksberg Palace; showing no interest in keeping these mound. Yet, both the burial mounds and the King's palace are part of a cultural history, which includes Carlsberg, that show a longue durée human engagement with this elevated topography. Seen in this cultural-historical perspective, Carlsberg is not isolated but rather closely related to regional urbanization processes. Carlsberg, then, is just as much a cultural-historical site as an industrial one.

The division of property shaping the urban landscape

Property lines and ownership has been a strong force in the formation of Carlsberg's landscape and has contributed to its spatial heterogeneity. The conglomerate that existed prior to the urban redevelopment suggests a structure of several independent areas. Yet, the property line at any time of this rapidly changing industrial site is but a snapshot in an ongoing process of disintegration and unification – a process that contemporary heritage, planning and design acts take an active part in, yet often in implicit ways.

Carlsberg is not simply one unit. Many companies have resided here simultaneously; at times five different enterprises related to brewing were doing business on the present Carlsberg premises. Various companies have developed as more or less independent enclaves, with their own manufacturing processes, power supplies, administration, road systems and storage spaces.

In the mid-1700s, the area of what would become the Carlsberg site was divided into two farms, both with fields oriented to the slope of the hill. Although property often changed hands quickly in the nineteenth century, the actual shapes of property areas that were apparent in the mid-1700s have had a remarkable permeability, its division affecting certain physical structures over the centuries. The former border between the farms has

37 From Rasmussen, *København*, 30. Translated by the author. Original quotation: 'en jævnere stigning og en lang kontur, som en stor ås'.

38 Larsen, *Gravhøje på Sjælland*, 145.

Image 2.11 Map from 1902 showing the New Carlsberg and the Old Carlsberg breweries asa two distinct systems. After the turn of the century these breweries had, however, begun to collaborate, for example with the new shared bottling plant Carlsberg Aftapningsanstalt.

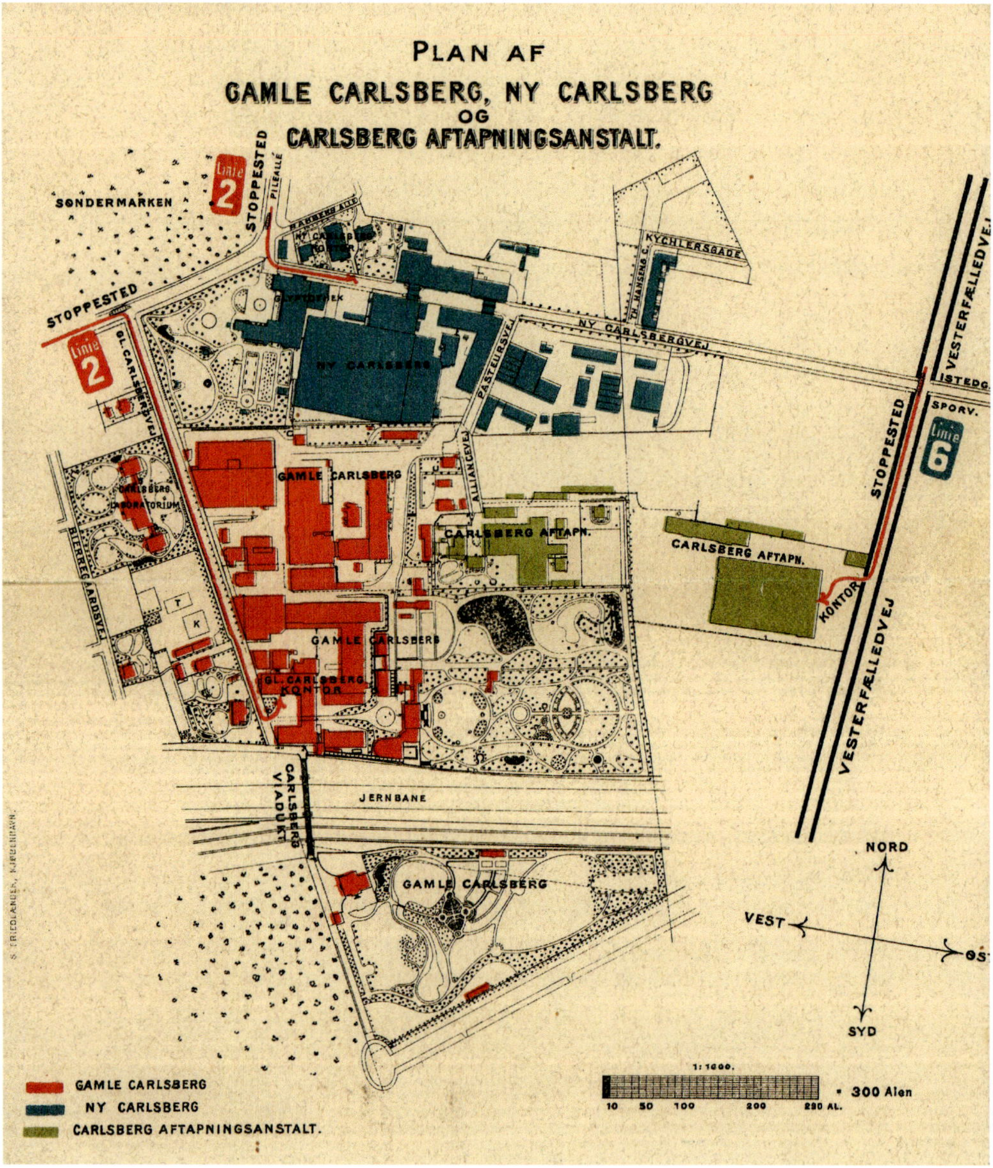

steered the later division of land, and can still be traced in the morphology of the site.

The farms gradually went fallow in the second half of the nineteenth century as a result of agricultural reforms and growing urbanization. The landscape was gradually overtaken by another type of production that could make use of the sloping terrain and rich aquifers in the area: breweries. In addition, other landowners started buying the fields, built housing on them and sold the land piece by piece.[39] During the late nineteenth century Carlsberg grew and expanded multiple times.

One event in this expansion has become famous for its epic drama and has long been an important element in the historiography of and the public perceptions of Carlsberg – and has even been the subject of a TV series on national television: the business rivalry between the founder of Carlsberg, J.C. Jacobsen (1811-1877) and his son, Carl Jacobsen (1842-1914).[40] The young and ambitious Carl Jacobsen – a flamboyant figure who as a boy had suffered from a stutter – was sent to work at different breweries in the United Kingdom to learn the business. His father's plan was that his son would lead his own brewery, next to his father's afterwards. J.C. Jacobsen built the new brewery and planned to rent it to his son during the first years of operation. While Carl was abroad, father and son exchanged vivid letters about brewing techniques and about the plans for the new brewery, which was scheduled for opening in 1871.[41] However, once Carl returned home, the hoped-for collaboration between the two collapsed and the father cancelled the son's rental contract. Carl then built his own brewery in 1880-1881 adjacent to the two others. This new enterprise, which he would later name New Carlsberg, grew quickly. The elder Jacobsen took back the brewery that he had first planned with his son – renaming it Mellembryggeriet (Intermediate Brewery), MB for short, in 1881.[42] The separation between those two companies and their competition in the years after 1881 is still legible in Carlsberg's urban landscape, just as later property divisions and building phases are. The image series on the next pages introduces to the main areas of Carlsberg in terms of spatial structures (2.12).

39 Most notably the wealthy businessmen F.C. Bülow and J.C. Ostenfeldt, who bought land from the Bakkegaarden and Bjerregaarden farms.

40 Glamann, *Bryggeren*, 230-255.

41 Glamann, *Din hengivne Jacobsen*.

42 This brewery was first called Albertina, then Valby Bryggeri, then Anneksbryggeriet and, from 1906 Mellembryggeriet (MB); see Zanker von Meyer, *Die Bauten von J.C. und Carl Jacobsen*, 51.

Image 2.12 Different areas at Carlsberg in terms of spatial structures prior to the urban redevelopment (2006). Dates showing when each area was built.

Old Carlsberg (1847-1860s)

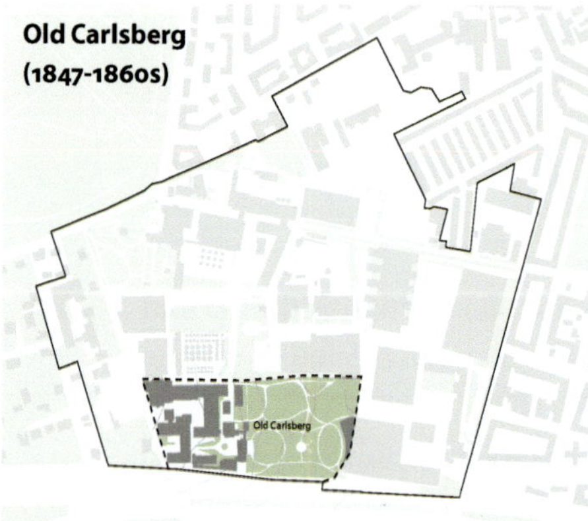

Old Carlsberg is organized along an east-west axis to the south and is a combination of a built-up ensemble and a landscape garden (J.C. Jacobsen's Garden). The scale and materials in the building complex are homogenous. The area is ordered into small terraces with meandering roads that descend sharply towards the east.

New Carlsberg (1880s-1901)

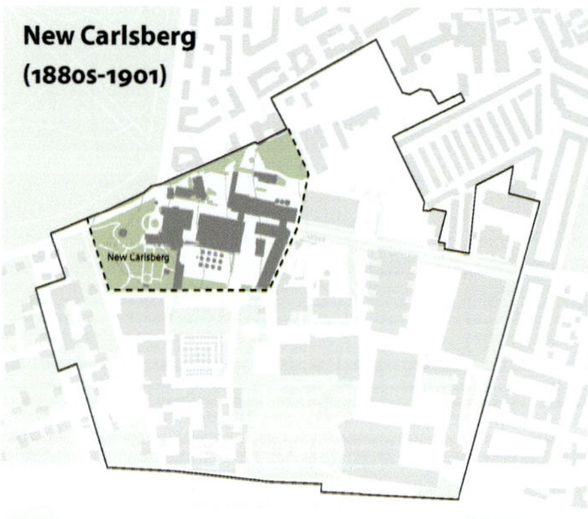

New Carlsberg is organized around the steep and narrow road Ny Carlsbergvej with buildings of varying sizes and shapes. It has a number of large, paved, sloping surfaces and a landscape garden that connects to the New Carlsberg Villa at the highest point to the west. Its buildings are rich in detail and were constructed with a variety of materials. Free-standing houses and gardens melt into the adjacent single-family housing quarter to the north.

MB Brewery (1860s-1870s)

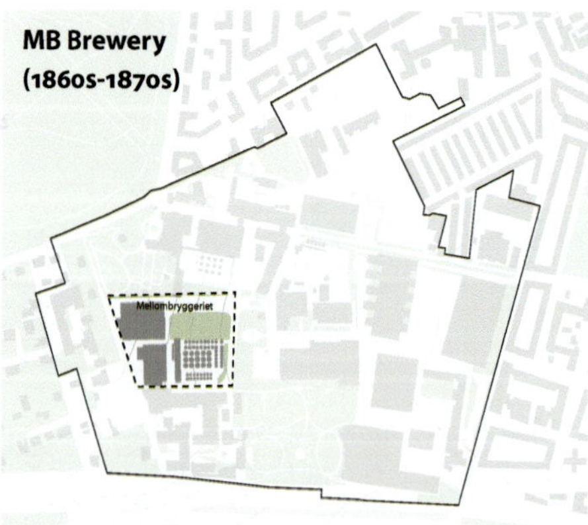

MB Brewery is immediately east of the Research Centre and north of Old Carlsberg consists of large buildings along a straight road. The terrain here differs substantially from the adjacent areas, because it has large terraces encircling a rectangular lawn, whose edges slope down sharply. A dominant element – a large cube of cylindrical steel tanks – was removed as part of the urban redevelopment.

Halmlageret, Maltsiloen and the KB Brewery to the north-east are characterized by large buildings and broad, open spaces paved with asphalt. A high fence marks a clear distinction from the adjacent attached housing of Humleby and the villas to the north. The terrain is relatively flat and a light-blue high-rise building stretches up along the Ny Carlsbergvej.

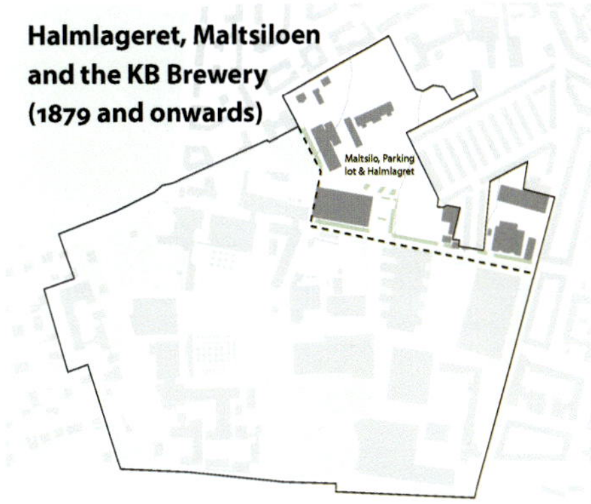

The Research Centre area to the west comprises free-standing buildings set in gardens, just like the adjacent single family housing quarter called Carlsbergkvarteret. A large ensemble of large buildings forms the Research Centre, with a symmetrically arranged garden in front of each building.

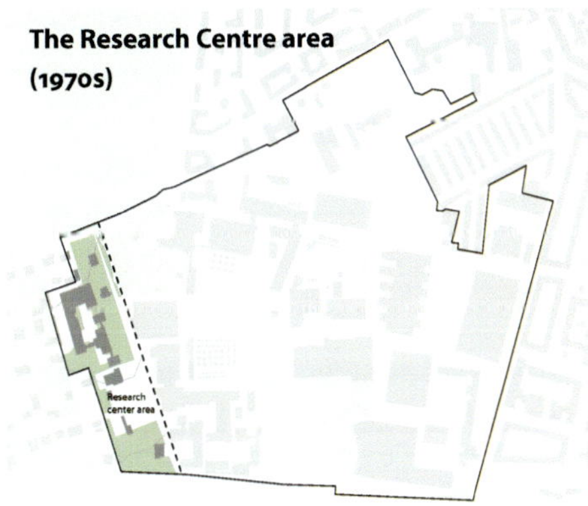

The Power Station and Airfield are at the centre of the Carlsberg site and posessed large buildings and a broad asphalt plateau supported by a planted slope and framed by chimneys and buildings. Three broad roads intersect close to a garden with an oval-shaped tree formation that mediates a dramatic terrain difference towards the higher MB Brewery.

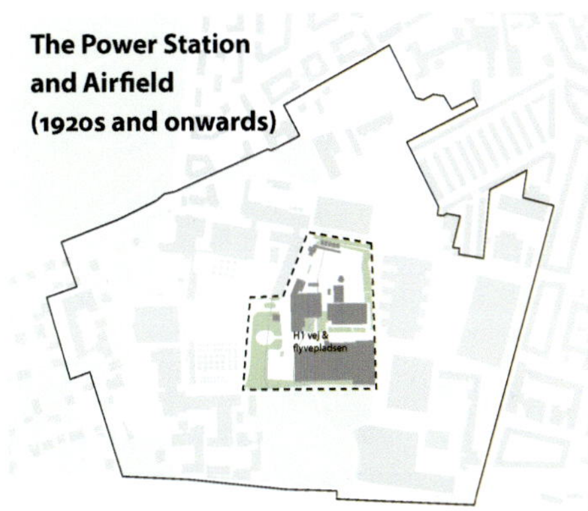

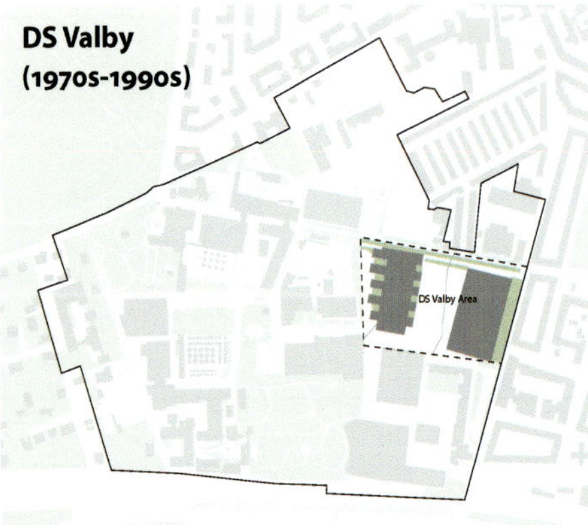

DS Valby was a large open space between two massive buildings on the eastern side of the site. The surface was almost flat and was paved with asphalt. A façade without windows and a steep slope face the adjacent neighbourhood Vesterbro, marking a clear separation from it. The large buildings are being torn down as part of the redevelopment project and the open spaces changed significantly as part of the urban redevelopment project.

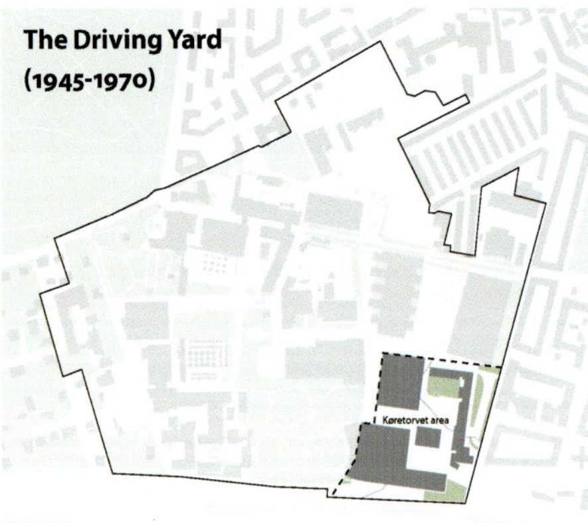

The Driving Yard to the south-east was the gateway to the city and clearly marked the brewery's separation from the adjacent neighbourhood Vesterbro. Large built masses framed broad spaces with asphalt surfaces and broad roads. A long brick building stretched along the road, but with a footprint that adjusted to the other Carlsberg buildings and not with the adjacent road. This part of the brewery was significantly changed during new construction in the first building phase of the urban project, begun in 2015.

Carl Jacobsen's New Carlsberg was built in opposition to the father's property to the south; its visual presence makes no concession to the older property in geometry, material, or scale. On the contrary, New Carlsberg's buildings are adorned demonstrations of power and wealth, and witness an exceptional passion for art and culture, which he shared with his father (2.13). In 1901 during one of the many expansions, the new Elephant's Gate (0.5) was built as a water tower and a symbolic entrance to Carlsberg.

Image 2.13 Ottilia and Carl Jacobsen with their children photographed in 1886 in their art collection, for which a special exhibition building had been erected at the New Carlsberg breweries. The brewer opened the collection open to the public and in 1897 inaugurated a larger sculpture museum, a glyptothek, close to the Copenhagen harbour.

Image 2.14 Beer label from the 1890s showing New Carlsberg with a perspective that makes Old Carlsberg (in the back) look smaller than it was. The swastica was the symbol for New Carlsberg, but the company stopped using it after 1945.

Image 2.15 Beer label depicting Old Carlsberg as a solitary enterprise with no indication of the adjacent New Carlsberg, with which it was competing. The large red building in the back right is from the MB Brewery that had become part of Old Carlsberg in 1881.

Old Carlsberg, on the other hand, today bears the evidence of J.C. Jacobsen's reaction to his son's oppositional decline of the father's offer. As Carl built his own brewery, the father created a new entrance that clearly distinguished it from New Carlsberg. This new and prominent entrance was clearly marked with the gate building (Kridttaarnet), which was adorned with Copenhagen's first buildings with electric lightning, and the gate (Stjerneporten). These buildings displayed the success and power by the Old Carlsberg breweries towards the son's new property. This entrance became iconic and was used in the marketing of Old Carlsberg and later by the merged Carlsberg breweries (2.16 left). The fact of the competition between the companies on the same site provides a backdrop for understanding the extensive ornamentation and the choices of self-representation in promotional materials, especially those dating from the time around New Carlsberg's establishment (2.14, 2.15, 2.16).

Image 2.16 Adverstisement posters from 1908 (left) and 1882 (right). Left: The significant gate building Kritdttaarnet was built to distinguish the Old Carlsberg Brewery from its then competetor, New Carlsberg. It is located at Pasteurvej. Right: The sculpture *Worshipper of the sun* symbolized New Carlsberg in the early years and is still standing on the roof of a malt silo building at the corner of Ny Carlsbergvej.

Danish author and journalist Herman Bang (1857-1912) visited Carlsberg in 1881, at the height of the conflict between the Old and New Carlsberg. He wrote an intensely personal report for the newspaper *Nationaltidende*. Bang was struck by the diversity of separate production units (including a bottling company called Alliance), which hindered the possibility of experiencing any sense of orientation:

> The thing is that Old Carlsberg and New Carlsberg and Alliance, to the uninitiated eye, can seem to be one, and yet there are three distinct businesses. The borders are, however, so difficult to discern that one moment you are on Alliance's property and the next on New Carlsberg's without knowing, as it meanders through such enclaves and bends that, as on a map of Germany with all its small principalities each with its own sovereign, the borders go back and forth. For a stranger, it is easy to get lost.[43]

Still today, the division of property is visible at Carlsberg and contributes to the impression of heterogeneity and confusion. The unification of property is however also legible; Old Carlsberg and New Carlsberg merged after both Jacobsens had passed away, beginning in the early 1900s. For example, shared facilities, a power station, an administration building, as well as cellar networks witness the merger.[44] In the twentieth century, the Carlsberg site went from being many companies to one process complex, which implied making new roads, corridors and other connected devices.

A landscape shaped by social order

Carlsberg's history is usually told from a company management perspective[45], and its architectural history has often brought this into dialogue with the intentions by the architects.[46] Yet what is less known is the spatio-political aspects of this landscape; how parts of it have been shaped to separate different social groups, mainly blue- and white-collar staff and the management. In the first decades of the breweries, J.C. Jacobsen was

43 Bang, 'Et Besøg paa Carlsberg.' Translated by Richard Hare. Original quotation: 'Man kan løbe vild herude. Tingen er, at Gammel-Karlsberg og Ny-Karlsberg og Alliance for almindelige Øjne løbe ude iet, og det er dog tre højt forskjellige Foretagenheder. Men Grænseskjellene ere saa svære at faa Kig paa, og snart staar man paa Alliances Grund og snart paa Ny-Karlsbergs, uden at man ved af det. Thi det gaar ud og ind med Enklaver og Bugtninger ligesom paa Tysklands Kort med alle de smaaa Fyrstendømmer, der hade hver sin Suveræn ligesom Karlsberg [...] Grænseskjellene derude gaar ud og ind, og for en fremmed er det let at forvilde sig.'

44 Jørgensen, 'Carlsbergs bebyggelseshistorie,' 54-55.

45 Glamann, *Bryggeren* ;Glamann, *Øl og Marmor.*

46 Keiding, ed. *Arkitekturen på Carlsberg*; Pedersen, *Bryggerens Akademi*; Vindum, 'Harrild på Carlsberg'; Frankel, *Gamle Carlsberg.*

understood as a patron, and many of the workers lived in the breweries.[47] Around 1870-1880 the working conditions became more regulated and the young director (Carl Jacobsen) introduced a more formalized welfare apparatus for his New Carlsberg workers than the senior Jacobsen had provided.[48] From the 1890s Carlsberg workers in both companies started to organize formally.[49] By 1900 the directors played a less paternal role in the lives of the workers, who now lived outside of the brewery and were on-site only during work hours. Women entered the production activities, primarily working in the bottling plants. In the second half of the twentieth century several demands from the workers led to improved conditions for them, such as reduction of noise, better working hours and mitigation of risks.

Two gardens exemplify how social order was articulated on-site. First, the landscape garden established by J.C. Jacobsen has a long history of excluding certain users. This garden has been part of the brewery since it was founded in 1847 and was planned exclusively for the use of Jacobsen, his family and guests in the Old Carlsberg Villa. This garden was clearly separated from the production area and the adjacent vegetable garden to the north by dense planting in the edges of the garden; gardeners and workers were separated from the director's family and guests. J.C. Jacobsen's garden was a quiet oasis and when visiting this arcadia you would not be able to see the everyday parts of the brewery with its production equipment and dirt.

The exclusiveness of the garden was reconfirmed in the mid-twentieth century during the introduction of a new bottlery building extension in the eastern part of J.C. Jacobsen's garden.[50] By this time, the villa was used as a retreat for professors and artists, and the new bottling station was requested to prevent the workers from seeing into the garden.[51] The bottling plant extension, as designed by the influential architect Svenn Eske Kristensen (1905-2000), complies with this brief in numerous ways. Excavated soil from the bottling plant's cellars was used to build an embankment that creates a visual border for the garden, planted with trees. The bottling plant's new façade is curved with windows to the sides (2.17, 2.19). In this way, the workers within the building had light but could not see the garden (2.18, 2.19). The exclusivity of the garden has been upheld right up to 2008. That year, a summer party for employees was held in J.C. Jacobsen's garden,

47 Ibid., 54.

48 Ibid., 60; Zanker von Meyer, *Die Bauten von J.C. und Carl Jacobsen*, 23-24.

49 Jørgensen, 'Carlsbergs bebyggelseshistorie,' 62.

50 Between 1967 and 1969, Carlsberg expanded the capacity of the Ny Tap bottling station by more than 30% (Markmann, 'Tegl i industribyggeri').

51 Kristensen, *Streiftog*; Kristensen, 'Udvidelse af tappehal for Carlsberg Bryggerierne.'

and later the same year, it opened to the public, except for the part of it closest to the villa. The garden is slowly becoming known to residents of Copenhagen, who increasingly use it as a public park.

Image 2.17 Geometrical analysis of the relationship between Old Carlsberg Villa (to the left on the drawing) and the bottling plant Ny Tap extension (to the right on the drawing), by Carlsberg Engineering Office and architect Svenn Eske Kristensen.

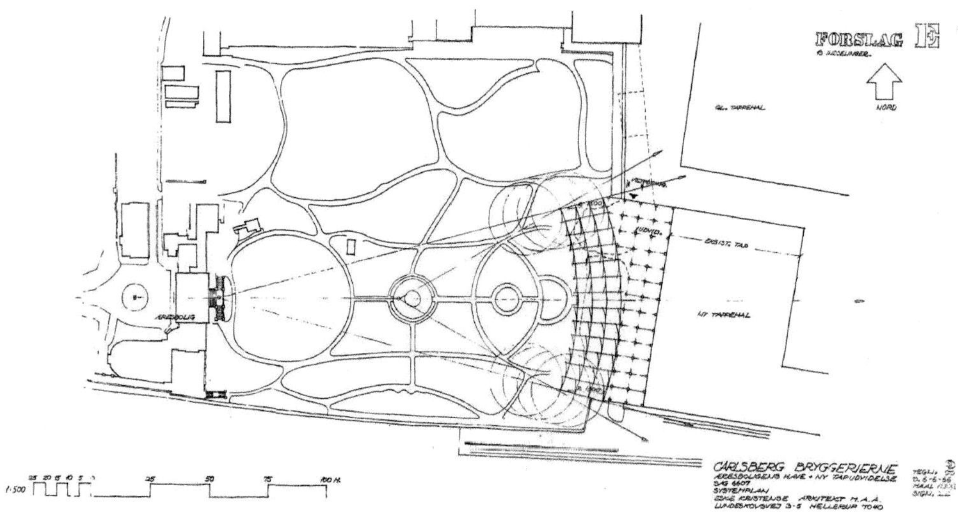

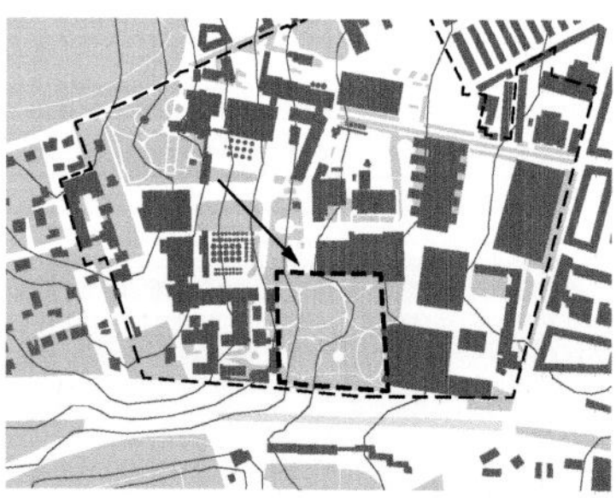

Location of J.C. Jacobsen's Garden

Image 2.18 Section of the Ny Tap bottling plant extension designed to ensure that employees in the bottling plant could not look into the secluded garden.

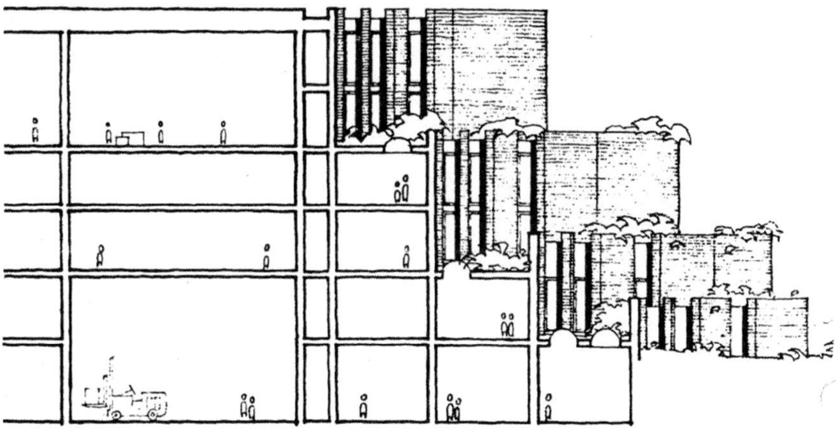

Image 2.19 The bottling plant Ny Tap's new façade towards the garden is sculptured in a way that prohibits people working within this building to look into the garden.

While J.C. Jacobsen's garden has been restricted to certain users, another garden motif at Carlsberg materializes other ideas about worker's social conditions. In the early 1980s a new bottling plant, Tap H1, was planned. The design brief contrasts with the Ny Tap extension of more than a decade earlier and rather relates to the strikes that Carlsberg's workers had carried out during the late 1970s to improve working conditions.[52] Also then a new bottling plant was built, but this time an impetus was to ensure a good working environment was important.[53] One of the novelties was to make several gardens so that employees inside would have a green view from the windows and green spaces in which to spend their breaks (2.20). This new plant, Tap H1, was a rectangular building with gardens that meander with the façade to the east and west with direct access from the building. The gardens, designed by landscape architect Inger Olesen, offered a variety of fruit trees and flowers.

In 2009, the Tap H1 building became a popular concert venue which also hosted congresses, festivals and exhibitions. To secure the transportation of equipment in and out of the building, and to separate paying guests to events in this building from others, the gardens were closed with gates that prevented public access and viewing. Paradoxically, the spaces that were once made to give more people access to gardens, were now secluded and public access denied.

Neither heritage surveys nor the plan for Carlsberg had given this bottling plant and its workers' gardens any special attention, and it was prepared for demolition (0.9). One proposal for the international ideas competition in 2006, however, by the design team SLA, had seen the gardens of the Tap H1 bottling plant as a quality for the future Carlsberg city district (2.21). While the designers seemed unaware of the social-historical significance of the gardens in the history of Carlsberg's workers, they were interested in the spatial composition that intertwined gardens with the building as a motif worth continuing and even exaggerating in the future city. SLA proposed to reverse the relationship between buildings and open space and to turn Carlsberg's large production and distribution buildings into publicly accessible enclosed gardens. The Tap H1 bottling plant thus became the material starting point for rethinking the relationship between inside and outside, garden and built-up space in the future urban district.

52 See for instance *Derfor kræver vi 6 timers arbejdsdag* and Jørgensen, 'Carlsbergs bebyggelseshistorie.'

53 Jørgensen, 'Carlsbergs bebyggelseshistorie,' 83-84.

Image 2.20 One of the gardens in Tap H1, depicted in 2008. Architect: Steen Højby Rasmussen and landscape architect Inger Olesen of Tap H1, built between 1982 and 1984 and demolished in 2016.

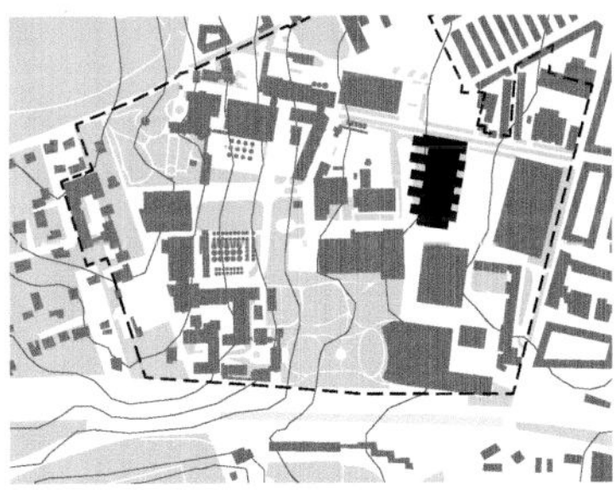

Location of Tap 1

Image 2.21 By removing the roof of Tap H1 and making it into a public garden, the design team SLA proposed to turn this part of Carlsberg into a publicly accessible mix of paved and planted urban spaces.

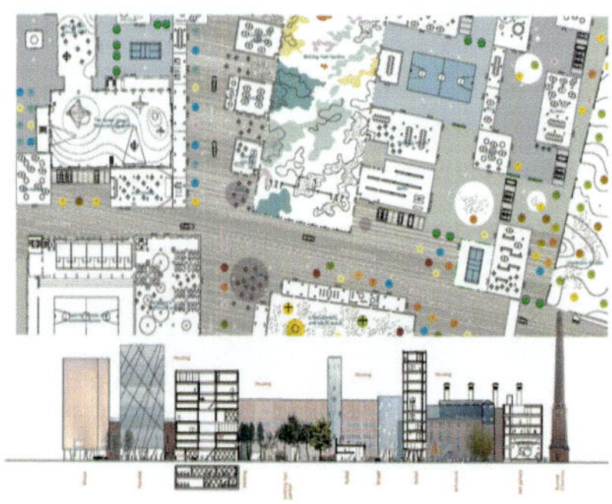

Concept

As these examples of gardens show, looking from a social perspective, Carlsberg is not one unified site, but reflects shifting social structures and changing relationships between unions and company management as well as between contemporary tenants and other current users. These gardens further underscore that by the act of surveying and proposing change –preservation, planning, design proposal –professionals, be it more or less explicitly and well-informed, take actively part in reconfiguring the social and material processes of separation and unification and establishment of hierarchies, insides and outsides.

Site definition: An interpretative and generative activity

The redevelopment of Carlsberg involved only an implicit and quite vague discussion about how the site is delimitated and why. Most competition proposals stayed within the property border, although some challenged it in their design. Although not explicitly argued, we can identify how the thinking from the discussions about place that were outlined in Chapter 1 came to frame the thinking and activities related to the Carlsberg urban site/heritage site, spanning between essential, constructivist and different relational approaches.

Many of the actors in the redevelopment were occupied with defining clear boundaries between what was inside and what was outside of Carlsberg's property line. This fixed perception of the borders served many means. First, in the heritage survey it seems to have fitted into the historiographic and public perceptions of Carlsberg as a (notably one, not many) brewery, and perhaps also into the administrative categories that would be easiest to use within the legal framework and established ways of working. Surveying Carlsberg as a unity that can be confined with a line on a map is simply the easiest way. Others who were occupied with drawing a sharp line between what is inside and outside of Carlsberg were the people who represented the corporate urban project. Their aim was to distinguish the Carlsberg urban project site as a brand and as an asset that could be parcelled into commodities on the market. Marketing the idea of a 'Carlsberg City' and making the new urban district into a clearly distinguishable urban fabric that differentiate it clearly from what is outside of it, all contributed to the need for branding the urban project. The knowledge produced in those surveys – particularly the heritage survey of Carlsberg's buildings – allows us to grasp some aspect of this site, namely, the fact that it has had gated borders to the rest of the city for decades and developed according to other rationales than the city outside of these gates. Yet, it only weakly grasps the multiple relationships that Carlsberg also engages in – as part of a larger urban landscape and of a transformative history that reaches further back than the industrial period and into a future where it can be entangled into the city in new ways.

Some of the designers and heritage managers were less reliant on the idea of unity and more interested in how the Carlsberg's connectivity. The designers behind the project 'Time Will Tell' refuse to fix Carlsberg in time and in space, as a unified entity with stable borders. Rather, it initiates a process in which there is no authorized place narrative, but rather an infinite process, in which all the people who will intervene in this site in the future participate (which then could become the place narrative). 'Time Will Tell,' then, seeks to initiate strengthened encounters between people, infrastructural flows and spatial situations at the borders. The project reinforces the porosity of Carlsberg's borders and thereby intends to speed up a process in which the formerly closed production site is now becoming more entangled with the city. If realized, the project would also be an architectural symbol of these relationships with highly visible infrastructural landmarks on most of the borders.

All the projects that are discussed in this chapter show different versions of what Carlsberg is as a site; they focus on different aspects of its Janus-faced relationship with the larger urban landscape once gated, yet always also entangled. Most of all, the difference between these projects demonstrate that sites are, as Kahn states *relational constructs* – not just neutrally confined, but constructed by *someone* to serve specific purposes.[54] As Andrea Kahn points out: 'Site construction posits that site boundaries shift in relation to the positioning – the physical location and ideological stance – of the beholder. It dispels the illusion of the city as either containable or controllable by hypothesizing the urban situation as a porous and shifting space.'[55]

Carlsberg can thus be defined in numerous ways because it engages in multiple relationships over time in prospect and retrospect. Cultural, economic, political and industrial processes of entanglement, unification and separation have contributed to shaping this urban landscape, as may every future intervention in these spaces.

The challenge of urban redevelopment processes then, is to stimulate a reflective discussion about the many possible sites and to prioritize on a substantial basis. Stakeholders in a redevelopment process can become

54 Kahn, 'Defining Urban Sites'. In Burns and Kan *Site Matters*, 294.
55 Ibid.

increasingly aware that the site is not a fixed unit but may just as easily be divided into many pieces, thereby providing an opening for the development of a future city with spatial and functional diversity, and potentially an openness to future changes and new layers to be added. Further, while the entanglement of formerly secluded industrial sites is a major urban challenge, understanding their relationship with the urban landscape is vital and can generate new answers to their future connectivity. Such questions concern designers and planners, but they also concern heritage professionals, who do not only engage with the past. Rather, these professionals (and many others) participate in urban redevelopment projects, including setting up the circumstances of its planning, programming, politics, uses and physical changes.

3 Space

Urban space is a major focus in Copenhagen's planning policies, which understands it as an important place to nurture people's well-being, social coherence and sense of community in the city.[1] Urban space also was a key concept and criteria of success in the Carlsberg competition and planning, which begun in 2006. Yet, while there were great expectations of Carlsberg's future urban spaces, the existing open spaces of the brewery went largely unnoticed and were substantially changed without much discussion. Surprisingly little attention was put into Carlsberg's outdoor spaces in the heritage surveys, design proposals and discussions surrounding the redevelopment process, although the undisputed stated ambition throughout has been to strengthen the existing character and history of the area. This chapter studies the redevelopment of Carlsberg as a locus for discussions of industrial open space and critically addresses the underlying assumptions and spatial concepts that were at stake in this discussion. From the observation that most of the existing open spaces of Carlsberg escaped from the valuation of most heritage professionals and designers, I add on new perspectives to the life history of Carlsberg's open spaces.

Urban space is an important locus for building a positive imaginary of Copenhagen – a city that has repeatedly been internationally awarded as attractive and 'liveable'.[2] The Carlsberg urban redevelopment project was an early a milestone in the city's urban space planning in the new millennium. The success of this as a 'vivid', 'attractive' urban district, would fail, the Carlsberg Ltd and the municipality stated, if they followed the standard procedure of planning and focused on constructing buildings.[3] Rather, the idea was to stimulate urban life by following an ethos that has since then become both a hallmark and a method in Copenhagen's spatial policies: when developing new city districts the planners aim to 'think urban life before thinking urban space and [...] think urban space before considering buildings'.[4] But what exactly does it mean to 'think urban

1 Copenhagen Municipality, 'Copenhagen Urban Space Action Plan'; Hermansen, 'Nye Byrum i København.' The main arguments in this chapter have been published in an earlier version in Riesto, 'Finding Industrial Space.'

2 Copenhagen Convention Bureau, 'Awards & Accolades for Copenhagen.'

3 Carlsberg Ltd Properties, 'Vores By Carlsberg.'

4 Copenhagen Municipality, *Arkitekturby København*, 58. Translated by the author. Original quotation: 'der tænkes byliv, før der tænkes byrum, og at der tænkes byrum, før der tænkes huse'.

space' in an abandoned industrial facility? How do different ideas about the value of urban spaces encounter the existing industrial landscape?

This encounter does not seem to have been neither desired nor easy to many of the designers who made plans for Carlsberg in 2006-07. Looking at all the 221 entries in the international ideas competition 'Carlsberg Vores By,' the jury wrote:

> [M]any entrants do not make any serious operational study of the syntax of the very different spaces around [...] [the historical] buildings or of the great spatial complexity and complexity of scale that characterize them. The 'industrial atmosphere' that all entrants seek to achieve can be found in these spaces and this complexity. It is the opinion of the jury that such a study would have added to the value of many entries – and that a study should definitely be conducted as part of any further design and planning.[5]

Such a study of the existing spatial structures in the former brewery was not made . Rather, heritage experts, designers and planners relied on much more implicit valuations of the existing open spaces on-site. Let us see how this works in a close-up study of one particular space that was valued in fundamentally different – yet often rather veiled – ways in the redevelopment: the so-called Carlsberg Square (3.1). Each act that has been intended to determine the future of this space – the competition brief, the urban plan and a later design proposal – addressed very different aspects of it, relying on different lenses and thought frames that were implicit, but nevertheless determined its future.

Reworking Carlsberg Square

The space now known as Carlsberg Square has one of the largest paved surfaces of the brewery, and it is partly enclosed by some of its tallest buildings.[6] The buildings close to the square are made of in different types of red bricks with different façade reliefs. The name was applied in 2008

5 Arkitektforening [Danish Association of Architects], Kvorning and Søholt, 'Carlsberg Vores By,' 24.

6 While none of the buildings surrounding Carlsberg Square have been removed, the eclosure has changed in the period 2006-2016 due to the removal of fences, trees and tanks.

after a public naming competition.[7] Previously, this space was called the Rinsing Yard, because this was where late-nineteenth-century workers had cleaned barrels – the first trading vessel.[8] This name has remained common in the daily language among employees at the brewery.

Image 3.1 The situation of Carlsberg Square in the breweries, as shown in 2005, just before the urban redevelopment planning.

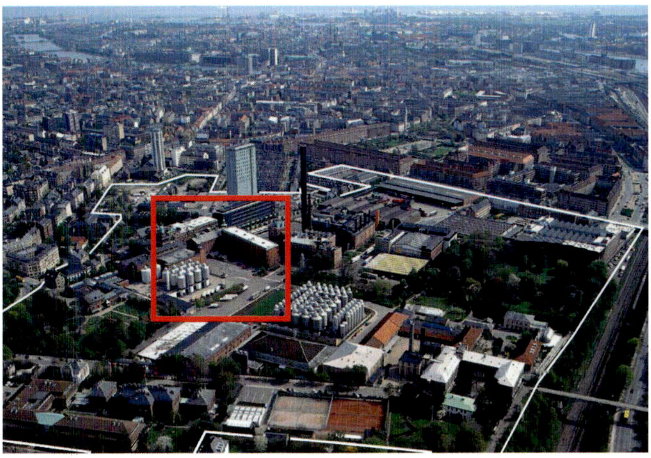

Prior to building the Rinsing Yard in 1881, Carl Jacobsen and his team had determined that wells could provide a suffiient supply for rinsing containers, filling the steam boilers, carrying out the brewing processes and watering the horses that powered distribution (until the truck took over in the 1950s). In the nineteenth century the Rinsing Yard was the main space of the then separate New Carlsberg brewery, and assembled all the core brewing functions: malting, brewing, fermenting, storing and distributing, along with power production and rinsing equipment. In 1906, however, barrels became redundant and a new bottling plant further down the hill became a new distribution hub. When numerous companies merged into the Carlsberg Breweries and a new central power station was built in 1922 the boiler houses on Carlsberg Square closed. The space from then on was used to park smaller vehicles and provide storage.

7 The Danish name is Bryggernes Plads (Brewers' Square). See Hjort, 'Carlsbergbyens Pladser er blevet døbt.'

8 The Danish name was Skyllegården.

Images 3.2 and 3.3 Carlsberg Square, and its axial connection with the Brewing House façade. Carlsberg Square mediates a terrain slope and is enclosed by large brick buildings.

Images 3.4 and 3.5 Carlsberg Square is arranged into two storeys by an approximately 2-metre retaining wall and two boiler houses.

Towards the north, Carlsberg Square appears as fully enclosed space with a large tarmac surface. The ornamental façade of the Brewery House is a focal point (3.2 and 3.3). Carlsberg Square has two elevations and a retaining wall roughly 2 metres high (3.4). Two small boiler houses, which produced energy for the industrial complex, stand at the conjunction of the two levels (3.5). The lower level gives direct access to a massive storage cellar, where horses could pick up goods and New Carlsberg's small train supply the boiler houses with coal. The dimensions of the space and its two storeys are major components of this industrial practice.

Images 3.6 and 3.7 Carlsberg Square has a tarmac paving with multiple repairs. Looking towards the south and west.

Towards the south and west, at the time of the ideas competition, the space was loosely enclosed by trees, steel tanks and containers (3.6, 3.7). Carlsberg Square has a trapezoid plan because it adjusts to the diagonal road that once supplied the Bjerregaarden farm. Its asphalt floor shows a pattern of repairs made to the top of the underlying cellar (3.6). The damage was caused by the cellar top reacting to temperature shifts over the years. The space now called Carlsberg Square is actually the deck of a beer storage cellar with exits on the lowest floor. This separated the black coal used for energy production from the rinsed barrels on the highest level of the space (3.8).

The practice of storing beer underground and the related practice of moving barrels in and out of the cellars determined how the terrain was modified in this space. The dimensions of the space and its two storeys are major components of industrial processes in New Carlsberg. During the redevelopment different actors have applied different spatial concepts,

Image 3.8 Drawing of Carlsberg Square from 1881, when it was called Rinsing Yard.

each of which emphasized open space, yet relying on different values and thereby perceiving Carlsberg Square in different ways.

Axes as open space value

In the competition brief, the Heritage Agency of Denmark and Carlsberg Ltd presented an inventory of the site's most 'distinct' and 'valuable' buildings and open spaces (3.9).[9] Why something is considered valuable, distinct, or neither, is not explicitly reasoned in the brief but certain principles can be deduced. If we look at the spaces selected there, Carlsberg Square is not presented as 'valuable'. In fact, only very few open spaces are selected. One of these is the space that is enclosed by the two famous gates – Diplyon and the Elephant Gate. This is part of the streets Ny Carlsbergvej and one of Carlsberg's, which have been publicly accessible during the entire time of production and are both the most prestigious and well-known spaces. The spaces that are given value in the brief are part of these publicly known parts of the site. Also, they are all enclosed by ornamented nineteenth-century buildings whose axes the space relates directly to. Open space, it appears, becomes valuable when it is clearly enclosed by adorned buildings from the nineteenth century and ordered according to an axis.

Carlsberg Square did not fit with these criteria – it has an irregular geometry, its enclosure is unclear, and, at the time of the survey, it lacked public significance. The space is, however, associated with value for another reason. The survey shows that an axis can be drawn from the ornamental façade of the Brew House through Carlsberg Square. Axes are the only spatial organization principle addressed in the survey, and all the axes found were, seemingly automatically, marked as valuable.

It must have been difficult to find axial spaces in this industrial process complex, and one may ask why axes were so important in the first place. It turns out that the search for Carlsberg's axes and enclosed spaces was prompted in a heritage survey fifteen years prior to the brief, in 1991.[10] Here, Carlsberg is assessed as part of the urban district Vesterbro, using the then new methodology SAVE (Survey of Architectural Values in the Environment). This internationally pioneering survey method developed along with the analyses of a broad range of Danish historical market towns

9 Carlsberg Ltd Properties, 'Vores By Carlsberg,' 31. Those categories echo the selection made in a heritage survey from 1991 but the inventory boiled down the number of valuable and distinct aspects compared to the previous survey. See Toft Jensen, *Vesterbro Bydelsatlas*.

10 Toft Jensen, *Vesterbro Bydelsatlas*, 15.

Image 3.9 Survey of Carlsberg's open spaces shown in the competition brief, with selection of the most 'valuable' and 'listed' open spaces. The inventory included 'valuable axes', 'valuable spaces' and one open space that had been listed part of Carl Jacobsen's garden. The red line is not part of the inventory, but shows the Carlsberg Square.

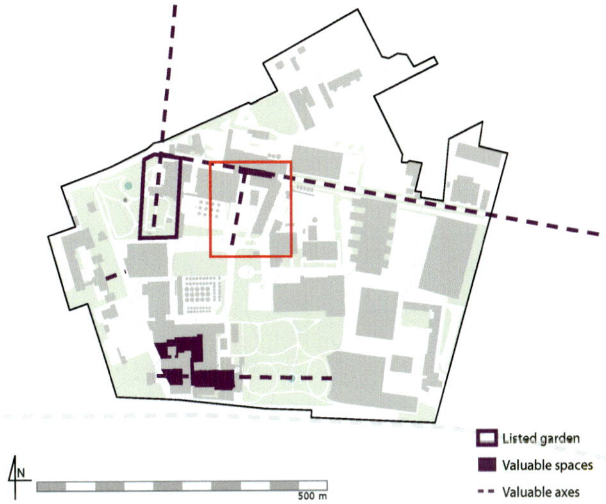

that were then high on the agenda.[11] Many of these towns have a central, enclosed square with an axial relationship to a monumental building. Findings in historical market towns appear to have influenced the scope of the Carlsberg survey, even though the official manual for the SAVE survey encourages an inclusive and reflective survey that is 'flexible enough to grasp the Palais of Amalienborg as well as the anonymous suburban single-family house but also a transformer station of a remote farmhouse'.[12] Although the aim of the SAVE method had been universal applicability, the SAVE survey of Carlsberg was in fact influenced by what was found in the places that were studied when the SAVE methodology was developed. As always happens, what was studied influenced how it was studied. In

11 Danish: 'Købstæder'. See, for example, *Roskilde – bevaringsværdier i byer og bygninger* and *Kommuneatlas Viborg*.

12 Algreen-Ussing, Silding, and De Wal, *Byens Træk*, 112. Translated by the author. Original quotation: 'at kunne indfange palæet i Amalienborg såvel som det anonyme parcelhus i forstaden, men også en transformatorstation på en afsides landejendom'.

the manual showing how to use SAVE, words like 'market place', 'square', 'axis' and 'focal point', which are all associated with pre-industrial urban forms, occur in the examples.[13] Further, the survey is not independent of being influenced by the people using it. The professionals who carried out the survey of Carlsberg did not necessarily adopt the refined, reflective approach the SAVE manual promotes but appear instead to have relied on a tacit practice that favoured certain open spaces in the often overly brief selection process.

The SAVE method was formulated as an embracing one that could combine topographical, historical and architectural perspectives of the built environment on multiple scales. But the SAVE analysis of Carlsberg shows that built objects were investigated in much greater detail than, for example, topographical and open space characteristics. Only a very narrow portion of the spatial characteristics of Carlsberg is emphasized, especially axial spaces surrounded by old buildings. These preconfigured ideals were not questioned when applied to an industrial landscape. The competition brief for 'Carlsberg Vores By' repeats the SAVE survey's valuation of a small number of 'valuable spaces'. This selection fitted well with Carlsberg Ltd's desire to minimize restrictions for the profitable urban redevelopment, and thus fuel the way for building densely. Emphasizing the open spaces with characteristic axes helped emphasizing the largest, most ornamented and prestigious old buildings that could continue Carlsberg Ltd's own narrative and uphold an established view upon aesthetics in the new district.

Pre-industrial spaces as universally good

Vores Rum' (Our Spaces) by Entasis (the winning proposal of the international ideas competition 'Carlsberg Vores By') argued that Carlsberg 'should be conceived of as space rather than as buildings' in order to encourage vivid and populated city life in the future.[14] The main idea was to transform the district of Carlsberg into a dense urban fabric where building blocks frame interesting urban spaces in different shapes and sizes that provide public life (3.10). Entasis proposed to make Carlsberg Square the most central urban square and to construct new buildings to the south and west so that it becomes enclosed to all sides.

13 In Danish: *torv, fixpunkt* and *plads*. Ministry of the Environment, *SAVE*. These examples have been changed in the SAVE manual from 2011: Stenak, *SAVE*.

14 See Entasis Architects, 'Vores Rum.' The quotation is taken from an elaboration of the project: Entasis Architects, 'Vores Rum: Plan Elaboration upon Competition Entry,' vol. 1, 13.

Image 3.10 Official plan for Carlsberg (lokalplan) 2008. The map shows existing buildings to be retained (brown), new 3-5 storey buildings (grey) and new high-rises (black). Carlsberg square marked with red, added by author.

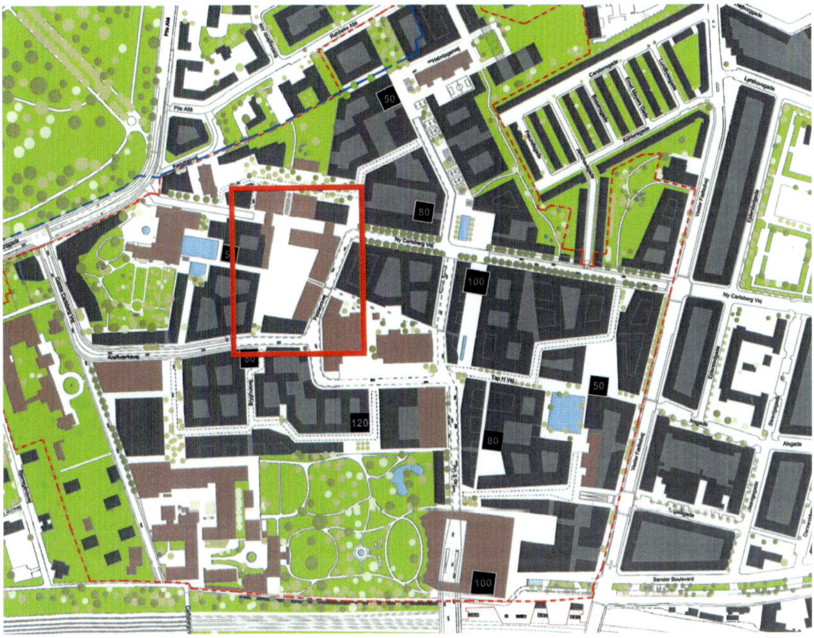

Why, then, did Entasis chose to change the district in this way? Entasis' competition entry included a diagram of two types of urban form: the 'modernist city', which foregrounds built objects in a continuous space, and the 'classical city', where spaces seem to be created out of a mass (3.11). In Entasis' view, the classical city is better suited to accompany human needs and ensure populated urban spaces with vibrant city life than the modern. The architects thereby repeat the dichotomy that was established by critics of modern city planning in the 1960s and 1970s, who argued that the modern city was unable to provide good urban spaces. Entasis' project relies on a specific understanding of valuable city life and this is closely related to one of the critics of modern city planning, Jan Gehl (3.12).

Architect Jan Gehl became known in the 1970s as a sharp critic of modern planning. Gehl argues that the squares of premodern European cities can encourage the vivid atmosphere that has been lost in many cities. Gehl introduces the concept city life (*byliv*) as a goal for urban public spaces. In Jan Gehl's widely read book *Livet mellem husene* (Life between buildings)

Image 3.11 Entasis' competition entry included a diagram of two types of urban form: the 'modernist city', which foregrounds built objects in a continuous space, and the 'classical city', where spaces seem to be created out of a mass.

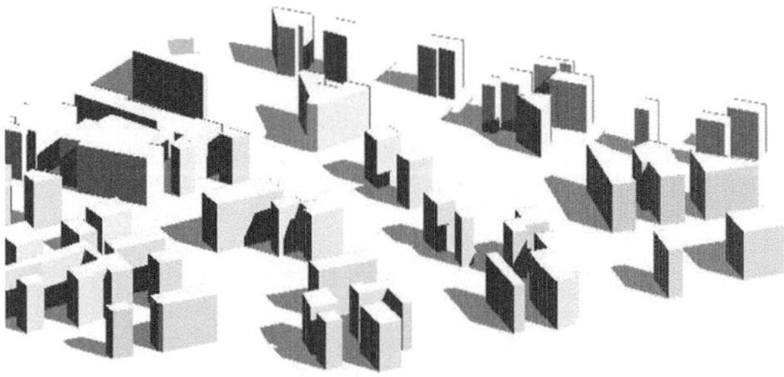

MODERNISMENS BYRUM

KONTRA

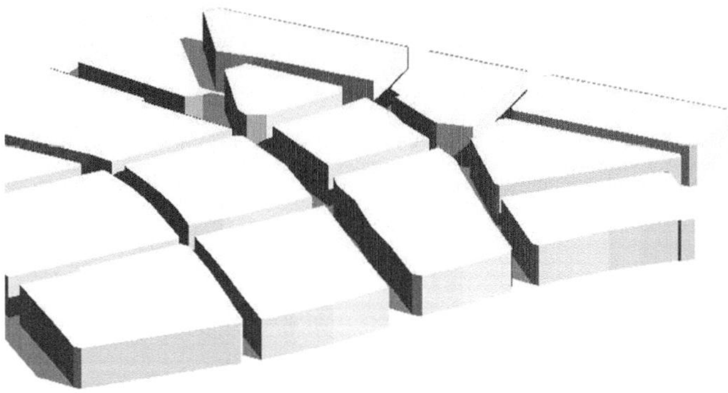

DET KLASSISKE BYRUM

from 1971, he finds the spaces of the modern city as ill-defined and inhumane, populated by free-standing objects and with weak opportunities for stimulating peoples' senses and social interaction. The classical European urban space, on the other hand, possesses a human scale and is enclosed by a dense urban fabric, and thus is better able to encourage sensuous stimulation, well-being and city life according to Gehl.[15] Gehl recommends aiming at a human scale with narrow streets and enclosed urban spaces

15 Gehl, *Livet mellem husene.*

Image 3.12 Pictures presented in Jan Gehl's book *Life Between Buildings*, where he criticized the modern city for being unable to meet basic human needs and instead encouraged his readers to learn from the open spaces of historical cities.

with a broad range of different shapes, façades that provide optical variation to a moving pedestrian, and many windows, doors, and niches. A certain urban form, so goes the argument, encourages certain forms of urban life – or even determines it.[16]

The term 'city life' as coined by Gehl was a core notion in the brief for the Carlsberg competition[17] and Gehl Architects advised the urban redevelopment process.[18] In the municipal plan, urban life is understood as

16 This is a widely criticized assumption. See, e.g., Sieverts, *Zwischenstadt*.

17 Carlsberg Ltd Properties, 'Vores By Carlsberg,' 40-41.

18 Gehl co-wrote the brief. A partner from Gehl architects was a competition juror and the office advised Carlsberg Ltd in the planning that followed after the competition bid. During the planning process subsequent to the competition Jan Gehl and employees from the office

populated spaces with, as the later detailing of the Carlsberg plan argues, 'classical urban space activities', such as 'shopping, outdoor serving, and events' (3.13).[19] Seen through Entasis' lens, which fits well with that of Gehl, Carlsberg Square shares qualities with the classical urban square: it is large, sunny, and (at least partly) enclosed by walls that provide optical variation. These aspects are believed to encourage the main goal: urban life.

In the further detailing of the competition proposal into a legal plan,[20] the boiler houses were prepared for demolition. Although from the 1800s and drawn by the architect of the Royal Theatre, Wilhelm Dahlerup, these boiler houses were not listed, because they had been rebuilt, which was seen as a damage to their authenticity.[21] A municipal appointment as buildings worthy of protection could have been possible but has not been the case. Instead, the municipal plan argues that these houses occupy Carlsberg Square's sunniest spot and restrict people's 'interaction with the buildings behind them'.[22] In addition, the space's two terrain levels are perceived as an obstacle to 'classical urban activities'. Consequently, the plan treats Carlsberg Square as a level surface and thus rejects the terrain difference because it is seen as a hindrance to other values.[23]

While the heritage survey had relied on an unspoken hierarchy that associated some aspects with value a priori, Entasis argues *why* some spaces are better than others, especially, their ability to encourage certain 'urban life activities', defined very specifically along Gehlian lines. Entasis' plan demonstrates a concern with history, not in the sense of the becoming of this specific location but as an evolution where the 'classical' European square peaks, declines and resurrects.

The urban plan by Entasis, instead, engaged with a dichotomy of two different types of urban form, in which Carlsberg Square can be seen as neither or both. It can be termed a 'classical space' because it is partly enclosed by buildings, even with ornamented façades that provide interesting sensorial experiences on a human scale. It can also be understood as a modern spatial

participated in several meetings with Carlsberg Ltd Properties and consulted the planning process formally and informally.

19 Copenhagen Municipality, 'Lokalplan 432, Carlsberg II,' 30.

20 Ibid.

21 Their interior has been replaced in the 1960s, and this made them unable to comply with the criterion for material authenticity in the practice of listing buildings. This argument was posed by the Heritage Agency of Denmark during the on-site workshop in July 2008.

22 Carlsberg Ltd Properties and The Heritage Agency of Denmark, 'Arven på Carlsberg , 6.

23 Copenhagen Municipality, 'Lokalplan 432, Carlsberg II,' 30. Gehl argues that small scale is important, which could have been an argument for retaining the boiler houses on the Carlsberg Square, but they were not part of the plan (Gehl, *Livet mellem husene*, 65).

Image 3.13 Diagram from the municipal plan (*lokalplan*) showing Carlsberg Square without the boiler houses. Instead, the space is understood as an arena for 'classical urban activities'. Red: zone for activities. Yellow: zone for lingering. Blue: zone for a water element.

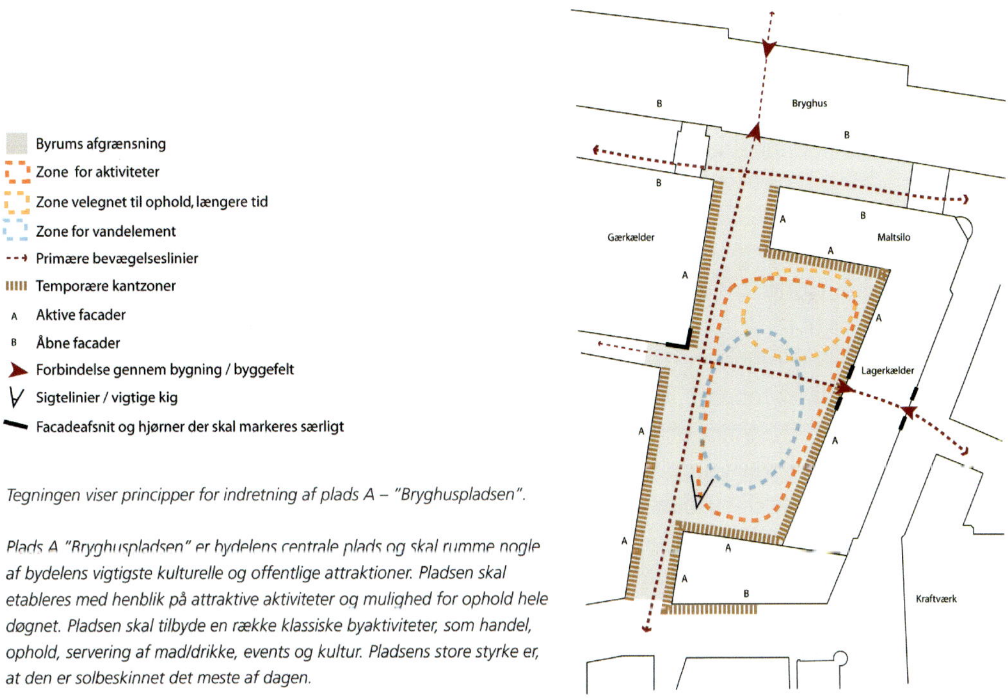

continuum with free-standing boiler houses that merges with the floating spaces of the brewery's road system. The characteristics of the Carlsberg Square as part of the tanker driving route's spatial continuum were, however, not articulated in any of the surveys discussed here. This conception of delimited spaces as valuable was set off in the competition brief and then repeated in the urban plan by Entasis while further reinforced by the new name Carlsberg Square, referring to historical city squares. Seeing Carlsberg through a lens that searches for enclosed spaces – squares – reveals a preoccupation with what is missing here rather than with what is apparent and imposes generic ideas onto the site rather than engaging in a dialogical relationship with it.

Appraising topography and landscape processes

When the international recession hit the Carlsberg urban project in 2008 and the realization of the urban plan paused, Carlsberg Ltd could no longer convincingly promote the idea of a populated Carlsberg with shops, cafés and other commercial outlets. Alternative stories and conceptions of the future urban spaces seemed necessary. In this situation, Carlsberg called in the Swiss firm Vogt Landscape Architects to add a layer to the urban plan, which was seen as required, anyway, so the recession years might as well be used for this detailed planning. Vogt was asked to provide an overall design concept for Carlsberg's future spaces and a detailed design for two of those, including Carlsberg Square.[24] The Swiss designers responded with a landscape plan that differentiates the site into three parts with distinct characteristics (3.14). These parts are neither indebted to the urban functions in the future district nor to historical periods. Instead, the conceptual division relies on three levels of Carlsberg's sloping terrain. Hence Vogt introduce topography as a core theme for the future urban district in a plan called 'The Story of the Hill'.

Vogt's design does not fight the existing two levels of the Carlsberg Square, as the urban plan did, but rather emphasizes them. The retaining wall is replaced by a large, S-shaped staircase with broad steps that allow for seating so that the intersection between the terrain levels becomes a meeting point (3.15 bottom left). The brewing cellars and the sloping, terraced topography have now become main aspects in the reading of Carlsberg's past, present and proposed future.

Seen through Vogt's lens, the small boiler houses are not obstacles to urban life activities but rather contributors to the space's attraction with its buildings on multiple scales, numerous histories and materialities.[25] The landscape architects do not see the bolier buildings as sources of information about the brewery's past but as components in a diverse urban space. It is thus not important to them to retain the exact materiality. In Vogt's design, only one of the boiler buildings is visible, yet radically transformed into a greenhouse with an entrance to the new underground parking area below. The urban plan had already determined that the Carlsberg Square should have a water element. Vogt resolves this requirement by letting

24 Vogt Landscape Architects, 'The Story of the Hill,' and Vogt Landscape Architects, 'Bryghuspladsen/The Carlsberg Square.'

25 Voss, personal communication. Voss was Vogt Landscape Architect's project manager for the Carlsberg commission.

Image 3.14 Main principle of Vogt Landscape Architects' landscape master plan for Carlsberg (2009-2010) called 'The Story of the Hill'. The designers define the site as distinguished into three levels that relate to the terrain slope, and the future open space design uses different paving, plants and occurrences of water on each level.

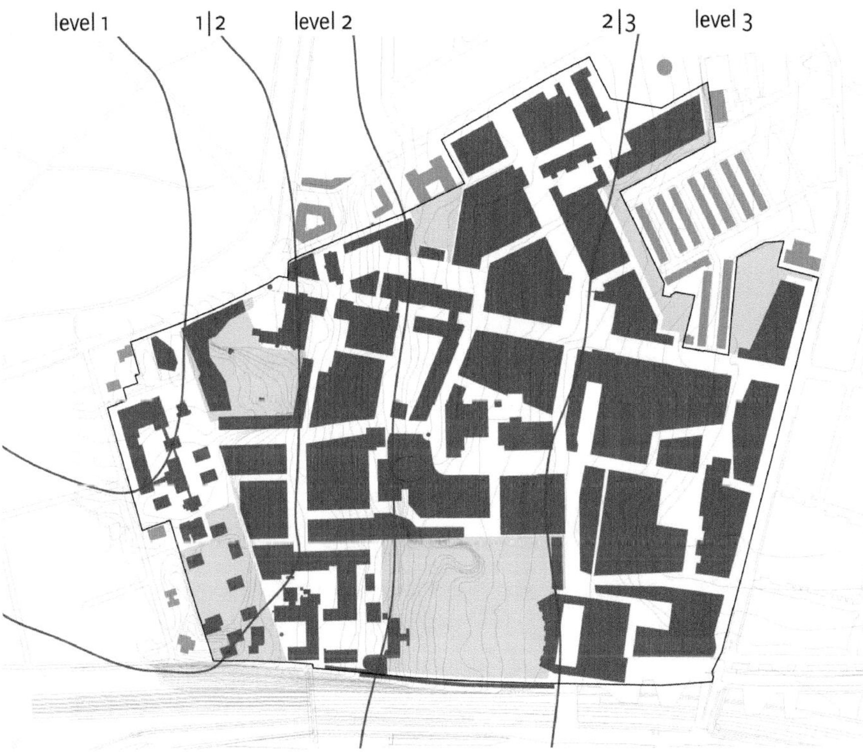

water appear differently on the space's two levels. On the highest level a pool collects rainwater and leads it through a subterranean filtration system. From here the water reappears on the lower level in a 'breathing fountain' (3.15 and 3.16), whose shifting water levels are steered through a control centre. This fountain of collected rainwater lets water occur on Carlsberg Square at different levels at different time periods. This proposal targets challenges of urban storm water and the need for water collection and recycling. This is not only solved technically but communicated in the design, which has a strong focus on narratives and symbols.

Image 3.15 Section and plan of the new Carlsberg Square according to the design proposal by Vogt.

Vogt do not understand Carlsberg Square as a separate urban form, as the heritage surveys and the urban plan had done. Rather, the space is now made up as an assemblage – which we can define together with urban studies researchers Ignacio Farias and Thomas Bender as a strong bond between 'the human and non-human aspects of cities – from nature to sociotechnical networks, to hybrid collectivises, physical artefacts and historical legacies, and the virtual or imagined city'.[26] In Vogt's project the assemblage is made up of soil and water, and different people's practices amid shifting cultural, economic and social situations in Carlsberg Square. Space, in Vogt's view, is a hub in a complex landscape-formation process – both retrospectively and prospectively. While inscribing itself in this transformative history, Vogt's survey, like Entasis', relies on specific thoughts about cities in order to synthesize and prioritize between different possible aspects of the space to guide the future development. In the book *Miniature and Panorama*, founding partner Günther Vogt is occupied with dissolving the binary of nature and culture and celebrates complexity: 'The potential of the urban landscape is the city's heterogeneity, a multiplication of the local.'[27] According to his view urban space must comply with the complexity of cities, and here landscape elements can contribute. Vogt calls for a city that offers enhanced experiences of sun, wind, rain and other changing landscape phenomena, and he emphasizes that this cannot be achieved by repeating formal solutions from the past.[28] Rather, Carlsberg Square's complex mix of different scales in the existing space is seen as a key to enhance specific

Image 3.16 The 'breathing fountain' of the proposed Carlsberg Square by Vogt Landscape Architects.

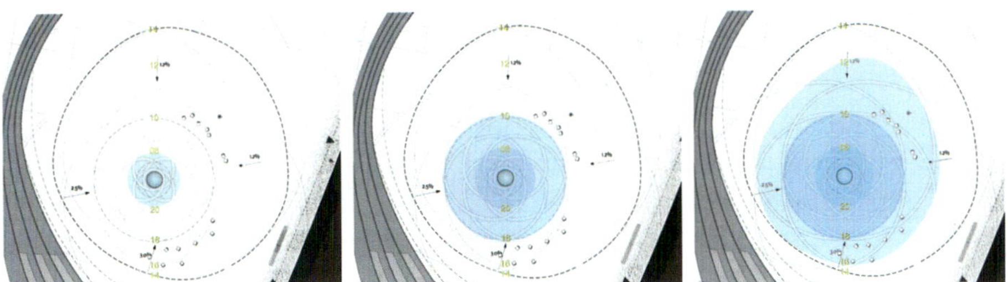

The extent of water at 08:00 and at 20:00 during the day | The extent of water at 10:00 and at 18:00 during the day | The extent of water at 12:00 and 16:00

26 Farias and Bender, *Urban Assemblages*.
27 Vogt, *Miniature and Panorama*, 101.
28 Ibid.

experiences. Vogt's design activates the existing landscape elements in order to enhance the bodily perception of future visitors – such as kinaesthetic experiences of walking upwards, spatial variation provided by buildings of different scales, a greenhouse with plants that reflect shifting seasons and the sound, smell and temperature of an ever-shifting fountain. Vogt relies on a radically different understanding of what is valuable city life than the Gehlian one presented by Entasis.

Vogt introduced a relational perspective on Carlsberg Square proposing that space is always in the making – and that its makers are both human and other agents in landscape processes. Vogt also stresses the relationships between social, cultural and environmental processes in this particular landscape and the act of design is both an intervention in those processes and, as is most prominent in this plan, a spatial representation of them. Vogt's plan illuminates an aspect that was weak or missing in the previous ways of surveying Carlsberg Square: topography. This was a determining factor in the industrial shaping of the space but it is ignored or even fought in the urban plan and missing in the heritage inventory. Understanding the effects of both water and the slope, however, reveals a new layer of understanding of Carlsberg's spaces pointing towards specific future possibilities.

At the same time, Vogt's project was also a tool for Carlsberg Ltd. It was designed when the financial crisis affected the site's redevelopment and the planned construction had been postponed. The goal expressed in the masterplan by Entasis, of a populated urban district with city life, in the sense of having shops and other commercial activities, had become difficult to sell to potential users and investors. Then, when Vogt's project introduced another vision, Carlsberg Ltd's agenda of making a profit was given a new positive narrative, because the main components of Vogt's proposal, natural processes in general and Valby Hill, in particular, are broadly associated positively. Vogt's project also very directly corresponds with Carlsberg Ltd's corporate storytelling to promote the urban project and the site. The abbreviated storyline that markets the plan sums up the history of the site: 'One man [J.C. Jacobsen]. One hill. One vision.'[29] This coming together of Carlsberg Ltd's economic interests in branding the urban project and the company with a heroic narrative and Vogt's ideas about a new role for nature in the city is important. It testifies that 'nature' is not a free zone, separate from politics and power, but is inextricably linked to human activity and interpretation, such as the narratives of an urban project.

29 Vogt Landscape Architects, 'The Story of the Hill.'

Chapters in the biography of Carlsberg's open spaces

Inspired by Vogt's design project and informed by Lefebvre's concept of space as the dynamic interplay between discourses, practices and materialities, I will now examine more specifically some of those relationships that have affected the formation of Carlsberg's open spaces. The following new histories presents Carlsberg's open spaces through a set of different lenses that complement each other and together make up caleidoscopic chapters into the life history of these hitherto often unseen spaces.

Yeast as an actor shaping Carlsberg

The sloping topography of Carlsberg is not always indicated on maps but is obvious to any visitor, walking upwards and downwards in a way that kinaesthetically contrasts with the way people usually move in the flat city of Copenhagen. Many spots on the Carlsberg site allow for lingering views towards the city and the terrain seems to unfold in many different ways: as terraces, as stairs, and as soft or steep slopes that are emphasized by towers, sculpture and high trees. The slope is seldom regular but separates into various terraces and levels, especially in the western, oldest and highest part. The eastern, lower part is relatively flat with many large open spaces created in the twentieth century.

This sloping terrain and the cellars dug into it are literally the foundation of the brewery's early success. Carlsberg led the market in the production of dark beer, which requires longer fermentation than top-fermented beers. This was a new product on the Danish market and its popularity was the key to the company's success.[30] During the second half of the 19th century, Carlsberg could not keep up with the demand for dark beer and could price it almost as they wished. But the costs of establishing a brewery for bottom-fermented beer were high. This kind of production required that the beer be stored cold and for a longer period than light beer. Maintaining the necessary zero degrees Celsius was difficult in the warmer months; the solution was to construct cellars that could keep the beer cold during fermentation and storage.[31] The slope of Valby Hill saved resources during

30 Dark beer became popular and between 1854 and 1871 the production of beer in general almost doubled, whilst that of the new and popular Bavarian beer increased tenfold (Glamann, *Bryggeren*, 76).

31 J.C. Jacobsen first tested this by constructing a cellar in the ramparts of Copenhagen in 1844. Three years later, he bought a piece of land from the Bjerregaarden farm on Valby Hill and started excavating cellars (Zanker von Meyer, *Die Bauten von J.C. und Carl Jacobsen*, 29-33).

the establishment of the brewery; the manpower needed to dig down vertically on a level ground is higher than digging horizontally into a slope. Herein lies the major economic argument for using this hilly site for brewery functions. The hill also contributed to efficient transportation of beer: the terrain could be modelled into terraces on which horses pulled barrels out of a cellar exit directly at road level.

In the early years the brewery introduced a new perception of Valby Hill by treating it as an economic asset. The practices of fermenting and storing beer underground and the related practice of moving barrels in and out of the cellars have determined the articulation of Carlsberg's terrain and turned Valby Hill into a slope of remarkable complexity on this site. Where adjacent houses and gardens have a relatively uniform terrain the brewery site has been modelled into terraces, of varying sizes, materials and shapes. As demand for bottom-fermented beer grew so did production and thus Carlsberg's cellars and the areas around them. When we compare the terraces in the first Carlsberg complex dating from the 1840s to the 1860s, with the second brewery J.C. Jacobsen built in the 1870s, the difference is striking. A drawing from 1889 shows the oldest brewery, which will later be called Old Carlsberg, set back behind many small terraces and three-storey buildings, while the second brewery, which will later be called MB, has four-storey buildings and fewer (albeit much larger) terraces over large cellars (3.17). The difference between the large terraces of MB and the smaller ones of Old Carlsberg is still legible in the terrain.

Numerous breweries with cellars have been established on the site that is now Carlsberg and were continuously extended in the nineteenth century. A large stone in front of Carlsberg's power station serves as a monument to all the effort put into excavating and altering the terrain here over 150 years. It bears an inscription that translates: 'This stone was taken out of Carlsberg's soil during the excavation of the ground for the storage cellar the year 1927.' Uses of fermenting and storing technology has had a large effect on the articulation of Carlsberg's landscape. The cellars were initially cooled with natural ice that could be stored for two years.[32]

The introduction of a new invention, a cooling machine, in 1878-1879 enabled new and more efficient ways than excavating to expand production. Now, brewing could take place all year round, even in warm summers, which optimized production tremendously. Further, storage could now take place in buildings above ground level, although they were still called cellars.[33]

32 Glamann, *Bryggeren*, 223.

33 Zanker von Meyer, *Die Bauten von J.C. und Carl Jacobsen*, 37.

Image 3.17 In Old Carlsberg (back left) the transitions between the terraces are less dramatic than at MB (front right) because the cellars are smaller. Drawing from 1889.

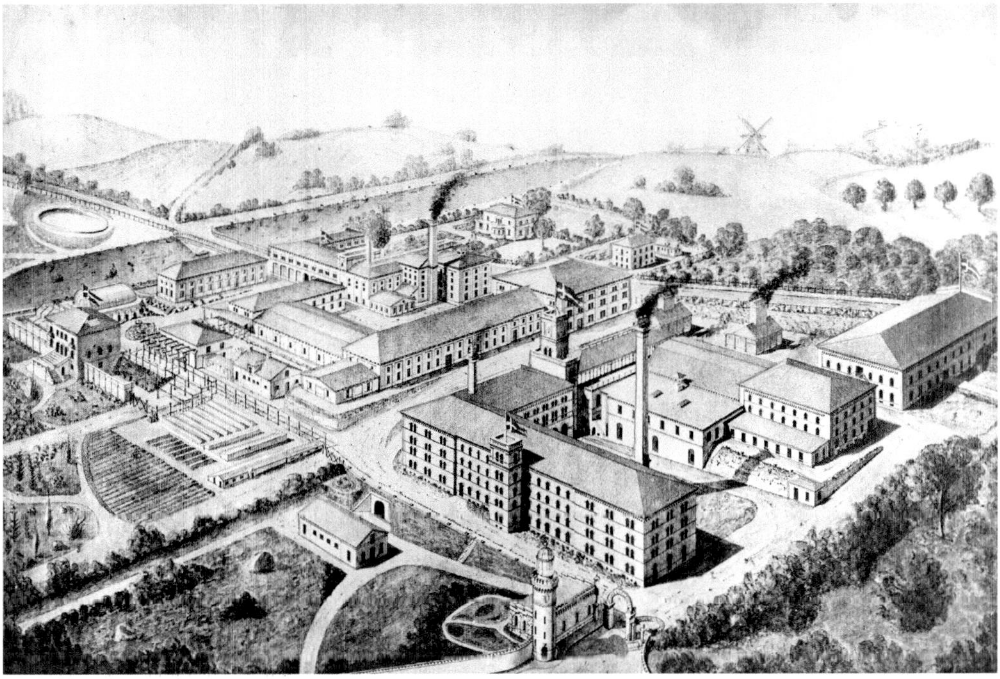

Storage and fermentation cellars in buildings that are above ground are much cheaper to establish than subterranean ones. Storage buildings generally were expensive; they had to cover a large area of ground to hold the horizontally laid large tanks in which beer was stored. A late example of a storage cellar above ground is Lagerkælderen from 1969 (See 1.2, back right). Because storage tanks could now be stabled above ground, and space and production optimized by building high, the size of this 'cellar' is substantial. The golden relief of the façade represent the horizontal beer tanks in this massive building, which Carlsberg's employees called 'The Castle'.

The latest chapter in the history of yeasting at Carlsberg came with a new type of rationalization in the 1980s. By this time fermenting and storage could take place in the same tanks, which were now vertical (3.18). These two novelties – vertical tanks and combining the fermenting and storage in one container – further optimized the amount of beer that could be produced on the same square metres, and made the logistics of getting it into tankers and trucks easy. Again, the new technology had to be integrated into the existing premises. Carlsberg Ltd chose to demolish most of MB and instal the vertical tanks on its large and plain terraces.

Image 3.18 Vertical metal tanks placed in a cubic form at the site of the former MB, following the plan of Henning Larsen Architects and Domus, 1993.

One of the last attempts to reorganize the Carlsberg site in Copenhagen as a brewery was made in 1993 (3.18-3.20, see also 2.2., 2.3). Carlsberg Ltd asked the renowned firm Henning Larsen Architects to investigate whether inserting more steel tanks (to expand production) was possible and to plan for a new headquarters building.[34] The plan that Henning Larsen Architects came up with in collaboration with Domus relied on a desire for symmetry and centrality. The plan aims at organizing and ordering the conglomerate. Old Carlsberg, MB, New Carlsberg and Tap H1 Vej are interconnected by a new central point – the headquarters building on top of the hill. The straight road that today leads up the hill to MB is turned into a descending garden that can be seen from the planned Headquarter building. Tanks and other new structures are designed in the geometrical forms that characterize what can be called postmodern architecture; for example, the cylindrical steel tanks are assembled into two cubes that flank the entrance to the headquarters.

34 Henning Larsen Architects and Domus Architects, 'Carlsberg Bryggerierne.'

These steel tanks were treasured by many architects in the international ideas competition who proposed various forms of reuse. Therefore, heritage experts discussed listing and preservation of selected tanks, but did not do it.[35] Today, all vertical beer tanks have been removed from the site.

Image 3.19 The plan by Henning Larsen Architects and Domus creates symmetry where two cube-shaped assemblies of steel tanks can be seen from the planned headquarters' garden façade.

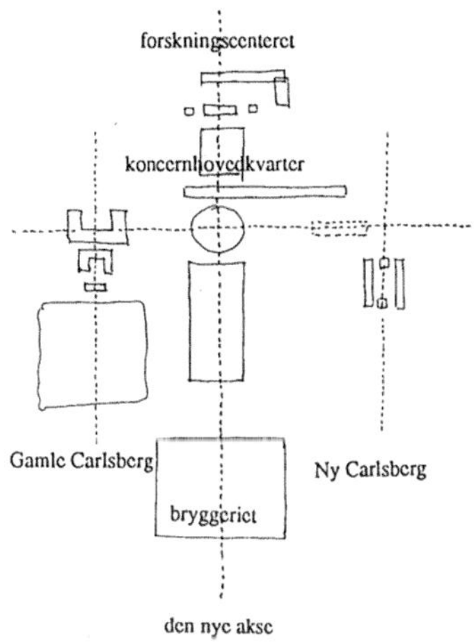

Image 3.20 The plan of Henning Larsen Architects and Domus for Carlsberg attempts to order the conglomerate in symmetrical syntaxes of geometric volumes with vertical steel tanks and buildings.

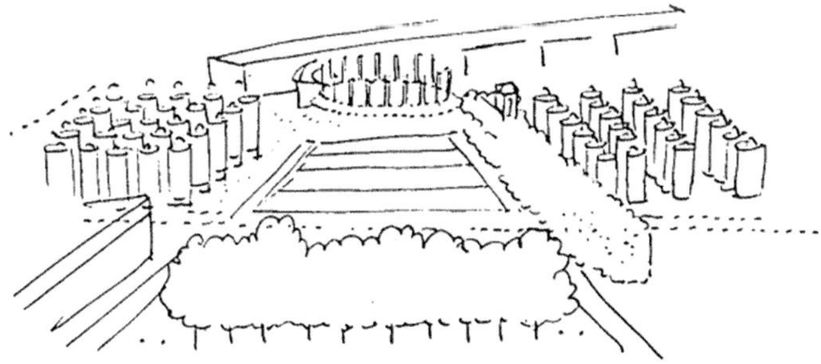

35 During an on-site workshop in 2008 with representatives from the Danish Agency of Cultural Heritage, Carlsberg Ltd Properties, Entasis Architects and the Copenhagen Municipality. I participated as an observer.

Aesthetics of landscape gardening

Whoever excavates a cellar is left with spare soil, sand or other materials. The massive and almost continuous excavating at Carlsberg, especially in the first five or six decades, has been accompanied by different ways of dealing with the spare soil, which have had an important impact on the terrain. The leftover material from the Carlsberg cellar diggings were first reused on-site but increasingly over the twentieth century were transported greater distances away. In 1847, before the buildings were erected, the spare soil from digging cellars was used to modify the topography while creating the director's villa and J.C. Jacobsen's garden.[36] The garden was planned in 1847 in cooperation between the owner J.C. Jacobsen and Rudolph Rothe (1802-1877), the Royal Gardener.[37] Rothe participated in the overall distribution of the entire complex that included buildings, a villa with a garden and the distribution roads.[38] It was unusual at that time for a production complex to be planned together with a landscape gardener. This first Carlsberg brewery was a narrow strip along the new railway line and descended down to the right. The buildings were placed close to the road on the western side, which had the steepest slope, so that cellars could be excavated. The residence fot the director and his family functioned as the junction between the higher brewrey and the lower garden.

J.C. Jacobsen's garden combines two spatial principles; a formal garden with an axial motif that follows the symmetry of the house and a landscape garden motif with soft lawns and meandering paths (3.21). The solitary trees have different shapes, colours and textures and belong to a wide variety of species. The ground is made up of pallet-shaped lawns, flower beds and sloping walkways. In the northern part of the garden the paths slope in a seemingly natural pattern, while in the southern part the paths are arranged as circles that adjust to the axis of the building.

The garden's topography is not naturally given but complies with aesthetic ideals of the time. The hill is shaped to provide the longest possible vista from the villa and across the broad lawn. To give the impression of the largest possible garden on the narrow site, Rothe later recorded in his notebook, he created a long view from the villa down towards the garden,

36 Glamann, *Bryggeren*, 60-61.

37 Jacobsen, 'Letter to Rudolph Rothe.'

38 The commission is to plan the environment at the brewery at Valby. See Rothe, 'Bemærkning til Planen.' On how this was an unusual way to plan a production complex, see Jørgensen, 'Carlsbergs bebyggelseshistorie,' 52.

Image 3.21 J.C. Jacobsen's garden in the first brewery complex, created in 1847 and onwards after plans by Rudolph Rothe. The garden combines two spatial principles. To the south is a formal garden with an axial motif that follows the symmetry of the house and stretches out with a circular planting bed and pond. To the north is a landscape garden motif with soft lawns that have solitary placed trees and meandering paths. The garden plan by landscape gardener Rudolph Rothe is lost; this drawing was made by E. Erstad-Jørgensen in 1913-1914.

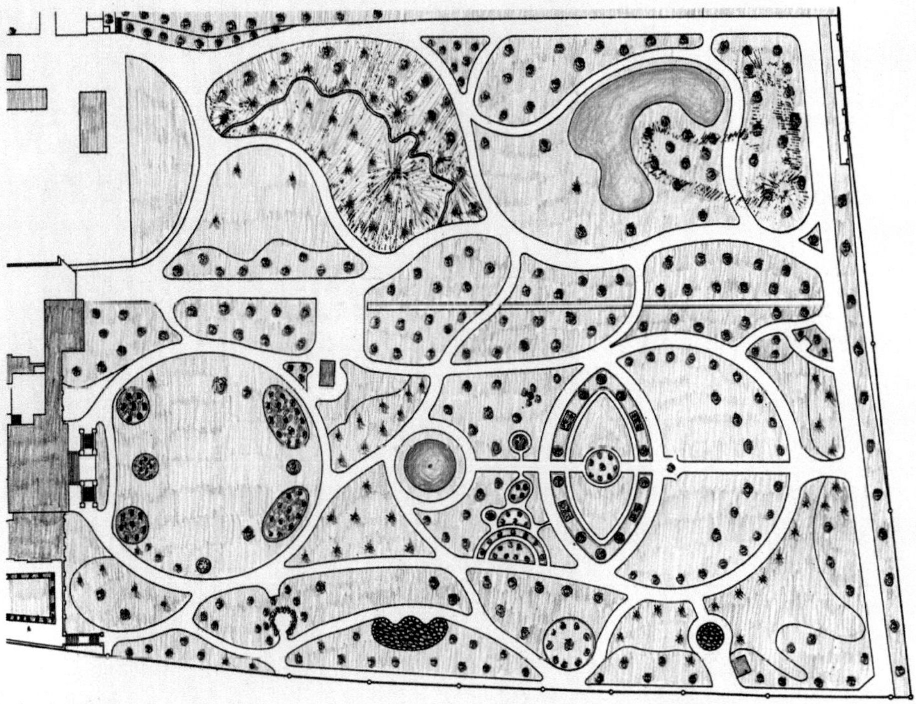

which was ordered 'relatively symmetrically'.[39] Soil from the industrial sites formed the base beneath the Old Carlsberg Villa, whose garden entrance thus became an impressive plateau. This way of planning the terrain is probably derived from a central source for inspiration in the mid-nineteenth century, John Claudius Loudon (1783- 1843), whose principles for villa

39 Rothe, 'Bemærkning til Planen.'

Image 3.22 Garden in Vienna drawn by Rothe on a study trip. It is arranged as a landscape garden with curved paths and trees and an open area with few trees in a straight line from the mansion, which give the lognest possible view down the garden. The edges are densely planted to exclude this arcadia from the world.

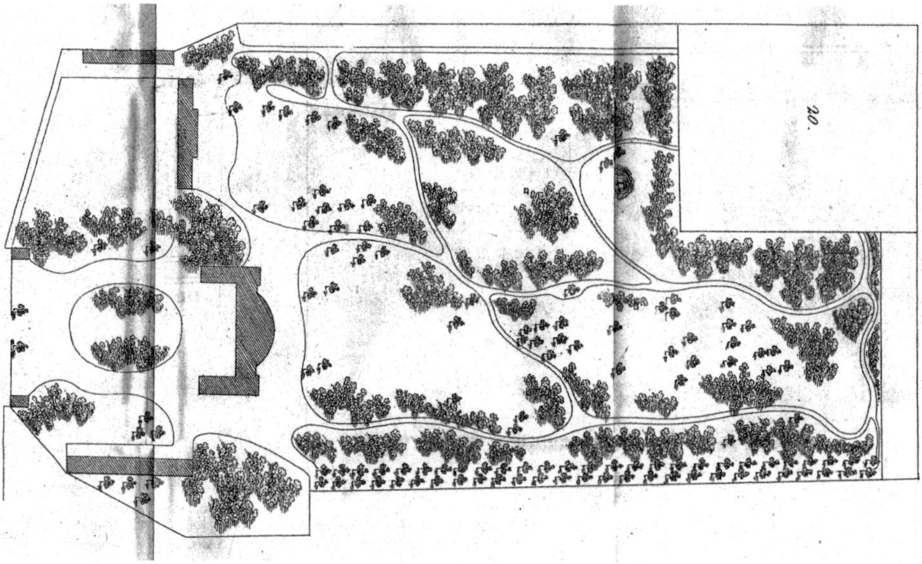

architecture informed new residence designs throughout Europe.[40] The Old Carlsberg Villa follows Loudon's principle of modelling a base which then becomes connected to the terrain on two levels: a high main entrance and a descending garden entrance on a lower floor. The garden entrance opens onto a monumental terrace and at the bottom of its staircase the terrain descends. This makes the villa look more prominent and allows for a distant view from the garden terrace.

The railway and the rise of international travel lead to increasing exchange of seeds and plants by gardeners in the nineteenth century.[41] J.C. Jacobsen's garden benefitted from this exchange and has many exotic trees. During a study trip in central Europe from 1824 to 1827, Rothe saw a garden by Fürst von Metternich in Vienna, which Rothe notes was an inspiration for how to work on a small site, his task at J.C. Jacobsen's garden (3.22).[42]

40 Loudon, *An Encyclopaedia of Cottage, Farm, and Villa Architecture and Furniture.*
41 Hauxner, *Fantasiens Have*, 72.
42 Rothe, *Udtog af en Dagbog Over Gartnerie.*

Image 3.23 J.C. Jacobsen's garden, a landscape garden with varied experiences of walking between trees with different shapes and colours.

The aesthetics of J.C. Jacobsen's garden's terrain provides a varied spatial experience. Rothe's garden is typical of those of the early 1800s, combining sloping roads with an open area with a long view from the villa. The garden of Fürst von Metternich and other landscape gardens have densely planted edges that foster a feeling of being separate from the surroundings with the paths arranged to achieve long walking routes. The aesthetics of the garden's terrain provides a varied spatial experience. It descends from the villa and downwards to the circular pond, which is surrounded by vegetation with soft edges. East of the pond is a homogeneous lawn surface which, although interrupted by flower beds, appears large and allows for long vistas because of its flat terrain. In the northern part two small and steep hills stand out from the rest of the garden. The path leads around them, while the hills are densely covered with shrubbery, needle trees and birches (3.23).

A few years after J.C. Jacobsen's garden was created, Rothe wrote: 'The more the terrain, when treated in landscape gardening terms, conforms to the laws of nature that have shaped the natural surface of the earth, the more its beauty is enhanced.'[43] In other words, the landscape gardener should not install completely new forms but adjust to and model the character of the existing landscape. This approach, Rothe argues, should also be strengthened in architecture to ensure the coherency of buildings and site. The landscape gardener and his internationally oriented client did not have difficulty agreeing on this approach. Years later, the brewer explicitly mentions the origin of their shared aesthetic ideal in a letter to his son: 'I will try to maintain my lawns in as beautiful a condition as the English.'[44] Rothe had visited England in 1841[45] and had read the work of garden theoretician Christian Cay Lorenz Hirschfeld, who was an important link between English landscape gardeners and the European continent. According to Hirschfeld, landscape gardens should not be installed as completely new forms but should enhance the naturally given property of the surroundings they are in. Seventy-four years before Rothe uttered similar ideas, Hirschfeld wrote about art as creating a new whole in existing landscapes, without deriving from what he called 'Nature'.[46]

In full accordance with Hirschfeld's and Rothe's ideas, J.C. Jacobsen's garden takes advantage of the existing hill and enhances it as an aesthetic principle.[47] Contemporary ideas about landscape gardening also inspired another way to reuse cellar soil. During a cellar extension in 1867, J.C. Jacobsen enthusiastically planned a new step in the development of his

43 Rothe, *Landskabsgartneriske Betragtninger over Danmark*, 19. Translated by Marion Frandsen. Original quotation: 'Jo mere der paa det Terrain, der i landskabsgartnerisk henseende behandles, finder Overenstemmelse Sted med Virkningen af de Naturlove, som have formet den frie Jordoverlade, desto mere forøges Skjønhedsindtrykket.'

44 J.C. Jacobsen, letter to Carl Jacobsen, 2 September 1869, in Glamann, *Din hengivne Jacobsen*, 274. Translated by the author. Original quotation: 'Jeg vil forsöge at holde mine Græsplæner i ligesaa smuk Stand som de engelske.'

45 Rothe, *Udtog af en Dagbog Over Gartnerie.*

46 Hirschfeld, *Theorie der Gartenkunst*, vol. 1. See, for instance, this quotation on page 145: 'Kunst bedeutet hier [...] die Schoenheiten, die sie in ihren Landschaften verstreuet, auf Einen Platz sammeln zu wissen; ein neues Ganzes, dem weder Harmonie noch Einheit fehlt, hervorzubringen; durch Verbindung und Anordung zu schaffen, und doch nicht von der Natur abzuweichen.'

47 The line of thinking can be traced back to the key figures of the English landscape garden William Kent, Lancelot 'Capability' Brown and Humphrey Repton, who had introduced the Arcadian as a motif in gardens. The ideas of the landscape garden were not about inserting perfect Poussinian images everywhere independently of the existing landscape but 'improving' what was there (Hunt, *Greater Perfections*, 32-75).

Image 3.24 Drawing by J.C. Jacobsen, who planned the hills of his garden. These hills combined soil reuse after cellar excavations in 1867 with aesthetic ideas about landscape gardening.

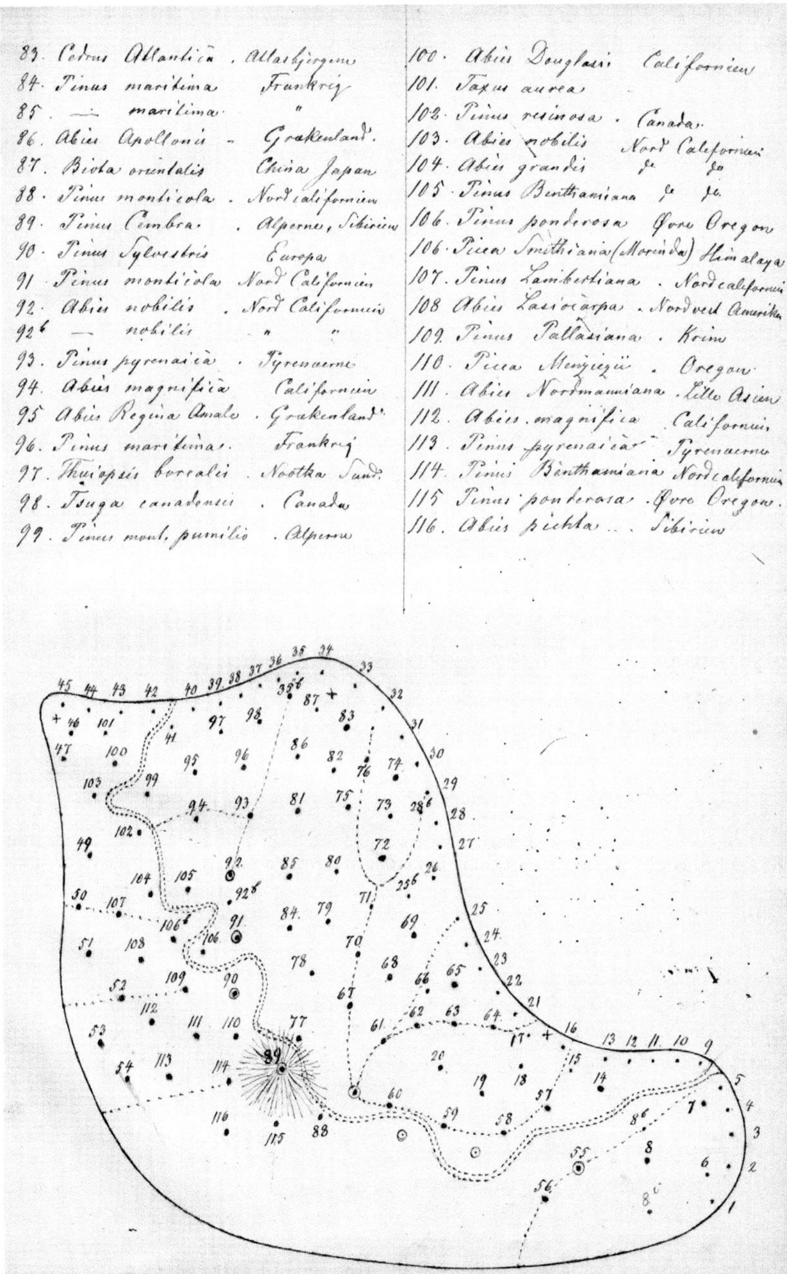

garden, this time without Rothe (3.24). The spare soil from this digging became the two small and steep hills on the north-eastern part of the garden, and in a letter to his son the brewer describes 'the use of the superfluous earth for the creation of the large "mountain" 30-34 feet [15-16 alen] in height in the new gardens [...] This "mountain" has recently occupied all my free time, as it is only through my constant supervision that aesthetic 'mountain contours' can be created, which I think has been achieved.[48]

The hill motif was important in many gardens in the late eighteenth and early nineteenth century. When planned on a site that already had dramatic topographical elements they were kept and 'improved'. But new 'mountains' could also be added, which became the solution at Carlsberg. During the extension in the 1860s the soil was not used to exaggerate existing terrain shapes but rather to install new and dramatic scenery for the walking visitor. These hills help to separate the garden into intimate spaces and allows for a varied walking experience through them. These hills also separate the garden from the production area to its north (behind the hill) and strengthen the impression of the garden as a little Arcadia, remote from the world of utility.

These hills are witnesses not only of how this garden is the result of a landscape gardener's ideas and expertise but also of how the owner as a brewer is a key shaper of his garden. As an industrial entrepreneur, Jacobsen combined industrial agendas with aesthetic and cultural ideals of his time, as the making of these hills show. These factors cannot be separated if we are to understand how the garden came to be this way. The current form of J.C. Jacobsen's garden is inextricably linked to the industrial rationale of the cellars. Yet, to frame it with established categories as either an industrial space or as a piece of garden art can easily blind us from discovering those interrelationships.

48 J.C. Jacobsen, letter to Carl Jacobsen, 20 April 1868, in Glamann, *Din hengivne Jacobsen*, 156. Translated by Marion Frandsen. Original quotation: 'Anvendelsen af den overflödige Jordmasse til Dannelsen af det store 'Bjerg' af 15-16 Alens Höide I det nye Anlæg [...] – Dette 'Bjerg' har I den sidste Tid beskjæftiget mig I alle mine Fritimer, da det kun er ved mit stadige Tilsyn at der kan tilveiebringes smukke 'Bjerglinier', hvilket jeg troer er lykkedes.'

Tanker route and social spaces

The largest open spaces of Carlsberg's hilly premises occupy ground that has been modelled into flat areas.[49] These surfaces are paved with asphalt or concrete stone. In the mid-twentieth century, when those spaces began to take on their current form, the brewery management began to perceive Valby Hill differently than in the nineteenth century: the hill was no longer considered an asset for production, as when subterranean cellars were needed. Instead, wheeled traffic necessitated large, flat surfaces. The increase in the use of cars and trucks to move goods and equipment between buildings, the continuing role of the railway station, the need for parking and the need to store containers and bottle crates all played a significant role in the shaping of the present open spaces at Carlsberg.

When the first Carlsberg complex was established in the nineteenth century, transport was powered by horse. Copper stone was the most efficient paving as it enabled a good grip for horseshoes. In the 1930s the distribution of Carlsberg's products was increasingly handled by motor transport, along with rail transport, which began in 1937. The most efficient paving for trucks was asphalt, which was gradually introduced around the plant, and flat surfaces created for parking. However, until the 1960s, horses were still used for some aspects of distribution; after that horses played a symbolic role in Carlsberg's marketing and they still tour Copenhagen occasionally. Since the 1960s Carlsberg's distribution has been a differentiated system of various types of motorized transport.

In 2008 I asked the workers I met on-site which open spaces had been given casual names during their daily use. The areas with names that they mentioned were the large open spaces paved with tarmac; each has had a central function in the twentieth-century distribution systems (3.25). These spaces also mark the distribution routes of tankers and other vehicles. Each name referred to either the brewery function, such as DS Valby, which is short for 'Distribution Centre Valby', or the social function of the spaces, such as the Airfield, which in the late twentieth century was where trucks parked after working hours and which formed a central meeting point and place for workers to have a chat after work (3.26). These large asphalt spaces, shown on images on the next page, were laid out for storage, filling of tankers and loading casks with mini-trucks.

49 For this investigation I am deeply grateful for the advice of and ongoing conversations with my colleague at the University of Copenhagen, Torben Dam.

Image 3.25 The route that tankers used to access the key points in the distribution from Carlsberg to the local market at the end of the twentieth century. Workers gave the large squares on the route casual names, especially the largest open spaces of the black marking; each has had a central function in the twentieth-century distribution systems.

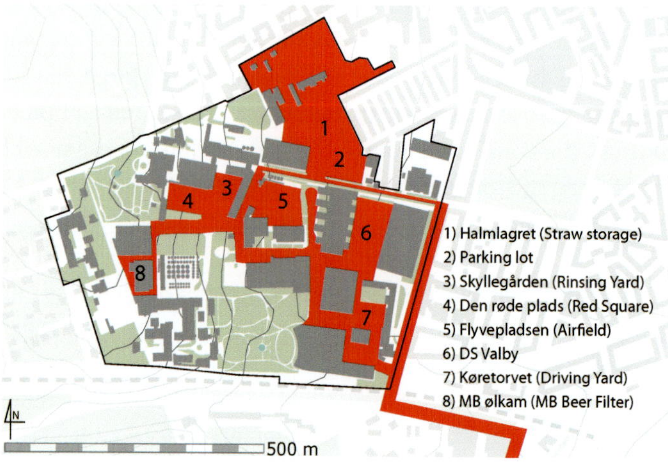

Image 3.26 The Airfield, a place to park the truck and meet up after working hours, parked as airplanes in an airfield. Scene from the 1970s.

Image 3.27 The Airfield, around 2008, as brewery production left most of the Carlsberg site. The space is being built at the time this book is published.

Image 3.28 Below the Airfield a slope with shrubbery mediated down to the large access area of Tap H1 Vej – a curved road that enabled trucks and tankers to turn. Both the upper and lower level are being built densely following the urban plan at the time of writing this book.

Image 3.29 Diagram of loggias and roofs in the brewery prior to urban redevelopment. Most are now demolished.

As the twentieth century progressed, more and more roads with curves (to cut the steepness of the incline) were needed, first for the trucks, and after the introduction of massive tankers. Installing such roads in the existing built fabric was often difficult because of a lack of space. This is one of the explanations for the many loggias that can be found alongside Carlsberg's roads (3.29, 3.30, 3.31). Most loggias are late-twentieth-century elements along the tanker route. Walking was very restricted, and the fragments of loggias, thermoplastic patterns and pavements, show where there were walking routes to canteens, meeting rooms and more. The loggias on the narrow streets used by large tankers separate the wheeled traffic and the pedestrians (3.30). These were part of the improved working conditions in the second half of the twentieth century. This latest chapter in the brewery's history, the trucks and tankers, have made a significant and expansive imprint on the landscape, adding a new layer over the terraces and the smaller copper stone roads that worked when horses were the main means of transport.

Image 3.30 Examples of loggias and roofs to protect pedestrians and goods in the breweries. Those loggias are today removed. Situation in 2008.

Image 3.31 Trucks at the soft drinks factory, Mineralvandsfabrikken, picking up goods protected by a roof. Detail of painting from the 1920s before the rail became an important means of transportation.

The tanker route that was installed in the late twentieth century has not been appreciated as worthy of preserving at any stages of the many recent heritage inventories, the international ideas competition or the urban planning of the site. Rather, most design projects in the competition perceive these large, open spaces as favourable grounds for placing new buildings. In the plan for Carlsberg diagrammatic drawings are used to show the future encounter between buildings and public space in order to make a vivid city. The plan does not refer to Carlsberg's existing spaces where buildings already meet the open space in differentiated ways with loggias, roofs, reliefs, niches, but prescribes that new buildings have open façades with many windows and doors in order to encourage an 'intense city life'.[50] According to the plan all new buildings must have public functions on ground level and the encounter between the interior and the open spaces is prescribed in detail. The differentiated encounter with open space among the existing buildings is not dealt with explicitly let alone considered valuable.

The open spaces along the tanker route, however, have become appreciated in other ways. During the 'waiting phase' following the financial crisis of 2008, when construction was put on hold, a charity car race used the precise affordances of the broad drive with its soft curves, as a driving route in central Copenhagen; DS Valby was rented by a go-kart enterprise. A temporary open space project, Ny Tap Plads, reused the roof that once covered goods on Kørepladsen between 2010 and 2015 (3.32, 3.34).

Creating large, level transportation and distribution spaces was a challenge for Carlsberg's twentieth-century planners. The possibilities for extending the property were limited in this conglomeration of more or less reusable buildings, on a hilly site surrounded by owners who were reluctant to sell. A common solution was to demolish existing buildings and replace them with flat surfaces. Often this had to be done on spaces where there were no cellars underneath, as those could not carry the massive tankers that drove beer out to suppliers in the city. After demolishing buildings and installing such plains, the incline of the hill was solved by slopes on the edges of the spaces, which were planted with trees, grass or shrubbery (3.35). The slopes are clearly defined with straight geometries and often planted with shrubbery or with plain grass surfaces. The green slopes might have been considered marginal in terms of transportation and storage but were appropriated in the everyday by the workers. Those sloping strips that connect plains on two levels became places to hang out and shortcuts in the daily walking routes.

50 Copenhagen Municipality, 'Lokalplan 432, Carlsberg II,' 3.

Images 3.32-3.34 Temporary installation 'The Rope Forest' by Keingart architects and UIW. By changing the Driving Yards' tanker deck, originally constructed to protect beer from sunlight during distribution, with ligtning and white ropes, new urban eperiences and activities are encouraged. Realized 2010, demolished 2015.

Image 3.35 Diagram showing the green slopes that resulted from terrain levelling in the twentieth century. Some of these slopes are currently being altered with the construction work during realization of the urban project.

Furthermore, the green slopes were meeting points. In contrast with the physical discipline a worker required on the assembly line, breaks in the outdoors afforded for another way to use the body than when working (3.36-3.39). The slope in front of the boiler house, for example, was a popular meeting place for female workers, who worked primarily in the bottling houses to the east and south of the slope, and male workers, from the power station to the north (3.37). Another such space was the so-called People's Park, which was situated along Tap H1 Vej (3.38). The hill might have been fought in terms of planning for transportation and storage but reappears as slopes of green with different functions, sizes, shapes and planting. This happened in a period when the amount and extent of the gardens at Carlsberg was radically reduced and more of the site became used for production, distribution and other rational functions. The slopes became green belts that allowed for everyday encounters and enriched the break times. The

Image 3.36 The controlled body along the assembly line. Women working in the bottling station, Ny Tap.

planting of those slopes with dense shrubbery and grass may have been a strategy to prevent such allowances. The green terrain strips allow for alternatives to the truck-dominated driving route. From the perspective of the people working in maintenance, walking to the lunch rooms or engaged in other blue-collar activities during the day, the hill was not an obstacle.

Image 3.37 This lawn is the intersection between the male working spaces in the machine centre and the female ones in the bottling house. A pleasant place to meet, such as on this day in the 1950s.

Image 3.38 A resting space in what workers used to call People's Park, which is between the power central and J.C. Jacobsen's garden. A sloping green space where people spent their breaks. This park was removed due to a brewery reorganization in the late twentieth century.

Image 3.39 The level storage and loading space of Carlsberg's former railway station Hof hosted parties and sports events, like this 'strong man' competition in the 1950s. As part of the redevelopment project, this location has been rebuilt into a new station, Carlsberg, which opened in 2016.

The new large distribution spaces, on the other hand, allowed for larger gatherings on special occasions. For example, a level storage and loading space at Carlsberg's former railway station, Hof, hosted parties and sports events (3.39) while employees danced in front of the Mineralvandsfabrikken (3.40).

The tanker route and large new buildings from the late twentieth century altered the terrain significantly. These changes, however, are not only the result of the formal, professional planning of the site to maximize its industrial role; changes to the site correspond with vernacular appropriation of the new open spaces and the green slopes between them, which were considered informal and intimate. These green slopes are often ignored and may to some seem like leftover spaces that are just the waste of the

Image 3.40 Space in front of Carlsberg's soft drink factory, Mineralvandsfabrikken. The size and paving of this space made it suitable for dance parties on special occasions.

plains. Nonetheless, they have already demonstrated their special potential as informal and out-of-doors meeting spaces in a differentiated system of spaces for various kinds of transportation, storage, social events and lingering. The spaces that had names among the workers were mainly those connected to certain functions in the process complex, such as storage, distribution, and transportation. Those spaces were striking during my first walk as an outsider, where they appeared as a succession of different atmospheres and experiences of motion. The current plan for Carlsberg contains no mention of the tanker route and no intentional spaces with a similar absence of programme for determined activities and meaning as the green slopes have been previously.

The first large building chapter of the new Carlsberg district, which opened to the public in 2016, applies an architectural language that seems to resemble contemporary international business architecture (3.41). It introduces new materialities and a new morphology to the site with little sign of what was there before. This is a quite different design strategy from the temporary urban space installations that had been at Carlsberg after the financial crisis. The conditions for these design projects were vey different. The brief for the temporary urban space projects was not to insert a new urban vision, as the brief in the international ideas competition for the entire site had been. Instead, the brief was to fill a five-year gap until the master plan could be realized. The designers of the temporary projects, Kienicke & Overgaard/Keingart, chose to reuse concrete physical structures on a small scale along the route, for example, the tanker roof on The Driving Yard (3.32-3.34), and the thermoplastic floor pattern along Ny Tap Plads. Their reuse did not arise from a study of the significance these physical elements had in the past, but on the potential for concrete reuse of physical structures to differentiate Carlsberg's urban spaces from those in the rest of the city and make them appealing. This physical reuse resembles Carlsberg Ltd's reuse of outdated technological facilities and building ornaments when it was still an active brewery. It is an example of how the new users produce new meanings. The aesthetic expression and use of these temporary urban spaces made them look like do-it-yourself (DIY) projects, in which users could concretely co-build the city and add their own layers of meaning and use without formal decision-making (see e.g. 0.8). Despite the visual similarity with DIY projects, however, those urban spaces were planned and financed by the private property development company with the objective to be a sign-post and generator for the economically driven urban project.[51]

Water as an actor shaping Carlsberg

Vogt Landscape Architects' plan suggests that hydrology is constitutive of Carlsberg's spaces, which will be the theme for this section. The site has a rich groundwater aquifer in a deep chalk layer. Geological movements in the Ice Age created the moraine of Valby Hill with fissures that directed groundwater.[52] However, the fault has also resulted in a sandy underground with a secondary water resource in the moraine layer that is not connected

51 Fabian and Samson, 'Claiming Participation.'

52 The following argument derives from Markussen and Kelstrup, 'Bryggerens Vand'; Glamann, *Bryggeren*, 64-65.

Image 3.41 Carlsberg's first large new building complex is a generic structure that does ot adapt significantly to the industrial landscape nor to the temporary uses of the site after 2008. This building complex opened in 2016 with educational facilities, housing, shops, restaurants and a new station called Carlsberg (at the site of the former station Hof). This is the part of the brewery that was formerly known as The Driving Yard and where the Rope Forest had been installed.

to the rich groundwater lower down. Underground pockets of water fill during periods of abundant rain, which, when completely filled, will sometimes even appear in the terrain above ground. Wells dug to reach those pockets might seem to be substantial sources of pure water, but they also could run dry quickly. This is what happened to the brewery's early wells, which were initially abundant but stopped supplying once a pocket was empty and so new wells had to be dug (3.42).[53]

53 This was the case in the first two decades of the Carlsberg breweries (Glamann, *Bryggeren*). Some Carlsberg wells contained water with large amounts of chalk, while 'soft water' from the

In the densely populated Copenhagen of the early 1800s sanitary installations were poor and pure water was a luxury. This is one reason for high beer consumption, since alcoholic beverages were safer to drink (they had a lower bacteria content) than water. While this created a favourable market for brewers, it was hard to find water that was suitable as a raw material for beer production. Water was important for mashing, which is the process whereby a mixture of hot water and crushed grain used to produce a malty liquid called wort. Another use of water was also used to cool down the beer during fermenting above ground. Water was also used in boiler houses that produced power in the nineteenth century and to clean equipment.

The assumption that Valby Hill would have ample water was a main reason that J.C. Jacobsen decided to establish his brewery here. Historian Kristof Glamann vividly describes the moment when Jacobsen visited the railway ditch as it was being dug in 1846 and saw water spurting out of the ground.[54] The wells his men dug did, however, turn out to be an unstable source. Lack of water was an ever-occurring threat to the survival of the breweries in the nineteenth century and the wells had to be redug numerous times. Natural sources were much cheaper than relying on the public water supply, so the brewers kept digging and searching. J.C. Jacobsen expressed the seriousness of the problem in 1869, while planning what is now known as MB: 'In relation to the location of the planned building project, I have strong scruples about placing it outside the utility and water supply of Copenhagen, as my deep well this spring has only just provided the necessary water for my brewery and malt house.'[55]

The first brewery on the hill, Old Carlsberg, was laid along an axis that went straight down the hill and through the villa. Every building in this complex and its garden adjust to the geometrical orientation of the villa, which was planned from the beginning. However, things did not go quite as expected. Out of desperation a well had to be dug in the new landscape garden in 1850 (3.42 well 4, 3.43). To draw enough water out of the deepest

municipal water supplies was considered more suitable for the 'normal Danish beer' in the late nineteenth century. See, for example, the letters from J.C. Jacobsen to Carl Jacobsen in 1869 about water quality, in Glamann, *Din hengivne Jacobsen*, 133, 279, and 281. See also Fraenkel, *Gamle Carlsberg*, 194-197.

54 Glamann, *Bryggeren*, 61.

55 J.C. Jacobsen, letter to Carl Jacobsen, 9 May 1869, in Glamann, *Din hengivne Jacobsen*, 266-267. Translated by Marion Frandsen. Original quotation: 'Hvad Beliggenhed af det projecterede Bryggeri angaaer, har jeg faaet stærke Skrupler ved at lægge det udenfor Kbhvns Grund og Vandforsyning, thi min dybe Brönd har I dette Foraar kun med Nöd og Neppe givet det nödvendige Vand til mit Bryggeri og Maltgjöderi.'

Image 3.42 This diagram shows the many wells that have been dug at Carlsberg during the first 50 years of production, when the search for water was almost constant. The year of construction is added, and it is shown which wells were there for the longest time.

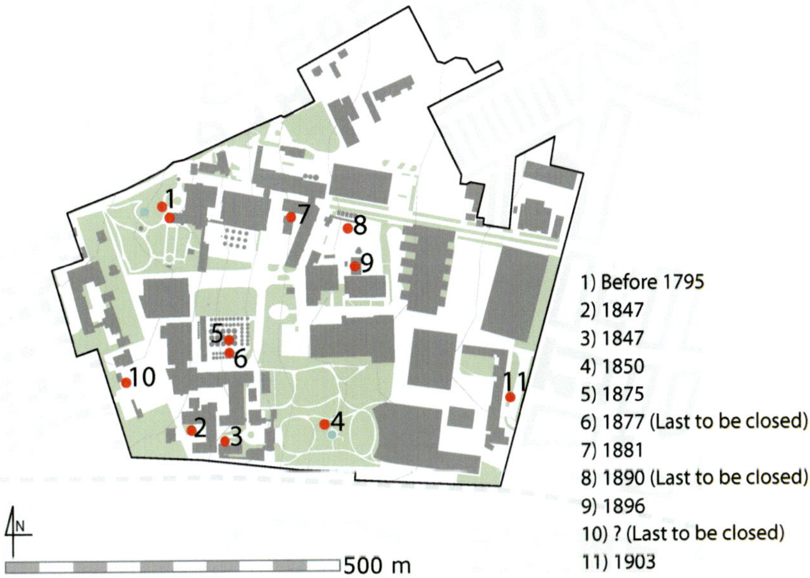

well a steam kettle had to be installed. Such an installation in the villa's garden went against the idea of separating utilities (vegetable garden, production facilities) from pleasure (the private garden). The steam kettle counters the symmetry of the garden. Jacobsen wrote:

> You can be assured that I just as reluctantly as you would disrupt the present arrangement of the development at Carlsberg, of which I am most content, as much for its suitability and convenience as for its beautiful symmetrical form, and it has caused me much trouble to find a way in which the new boiler could be placed [...] without altering a part of the site and thereby disrupt its cohesiveness and symmetry.[56]

56 J.C. Jacobsen, letter to Carl Jacobsen, 3 July 1865, in Glamann, *Din hengivne Jacobsen*, 31. Translated by Marion Frandsen. Original quotation: 'Du kan troe, at jeg ligesaa nödigt som Du ville forstyrre det nuværende Arrangement af Bryggeriet paa Carlsberg, hvormed jeg selv er særdeles tilfreds, saavel for dets Hensigtsmæssighed og Beqvemmelighed som for dets smukke

Image 3.43 The steam engine that had to be placed in Carlsberg's garden (today called J.C. Jacobsen's garden) quite contrary to the idea of separating utilities (vegetable garden, production facilities) from pleasure (the private garden). Black line added to show the axial disposition of the built complex and the new well with the steam engine (black circle, added to a drawing from 1897).

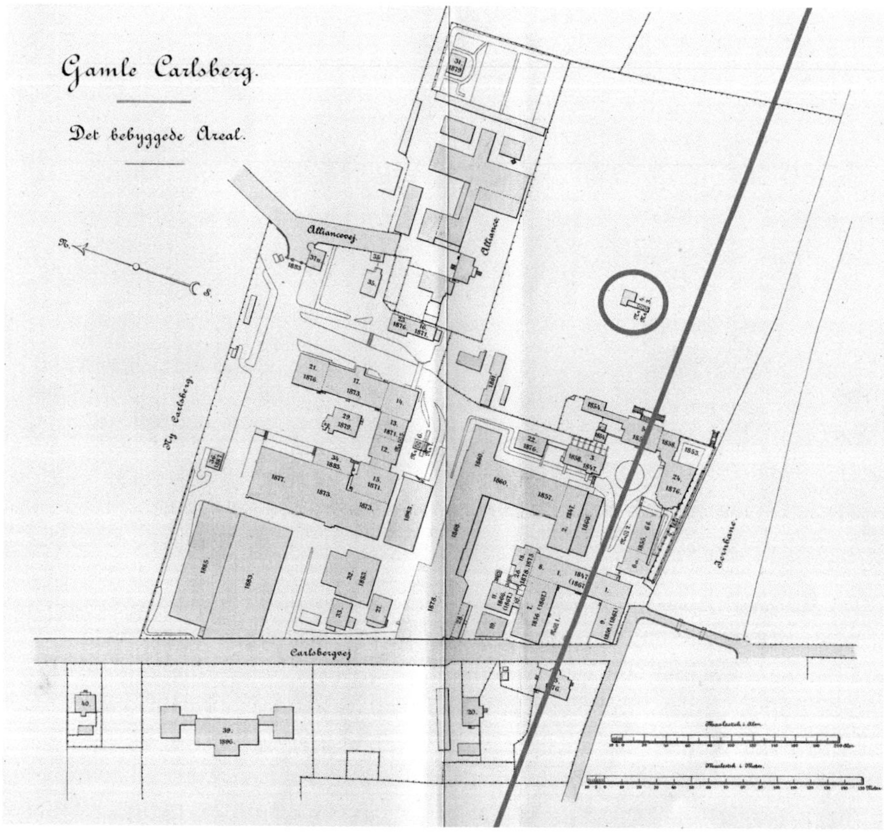

The very look of the huge engine as well as the continuous transport to and from it was definitely not what was planned for the landscape garden at the director's villa. The continuous and often dramatic search for water determined the directions in which the brewery extended and how. The

symmetriske Form og det har voldt mig meget Hovedbrud, at finde paa en Maade hvorpaa den nye Dampkjedel kunne anbringes og bekvemt sættes i Forbindelse med Mæskekar og Mæskekjedel uden at omforme en Deel af Bryggeriet og derved forstyrre dets Sammenhæng og Symmetri.'

hydrological circumstances had a complex relation to the planning of the Carlsberg breweries in the first decades. On the one hand, the water was a main reason for the location of the three breweries, and, on the other, it was a primary force in the changes made throughout their time of production. Sometimes architectural ideas had to be given up because the water supply was empty and a building's role had to change. After the 1870s new pumps had enabled direct access to the groundwater, and the crucial water supply became more reliable. The Carlsberg area is today connected to the water supply of Copenhagen and of the neighbouring municipality, Frederiksberg.[57]

Industrial open space as multidimensional lifeworlds

While the existing open spaces were surprisingly absent in the discussion of the Carlsberg redevelopment, some of the most iconic spaces were, in fact, examined and debated. The heritage experts who surveyed the Carlsberg site relied on a canonical understanding of open space that saw certain forms as better than others. Similarly, the designers relied on multiple tacit and explicit assumptions about why some spaces are better than others, and according to this understanding the existing brewery complex largely failed to comply with the norms. Moving from those quite narrow normative perspectives, my attempt in this chapter has been to build up alternative knowledge about Carlsberg's existing open spaces and to test a working method that relies on another conception of open space. Rather than to identify the forms as essentially good or bad, my attempt has been to show Carlsberg's open spaces as relational, and as the interim outcome of a dynamic interaction between social, cultural and environmental processes that has created particular appearances, spaces and possibilities for bodily experiences and interpretation over time.

The hilly topography of Carlsberg has been utterly defining for how this site has transformed from the time people first walked up it to worship the sun in the Bronze Age. The hill has been celebrated and associated with aesthetic, cultural and, beginning with the brewery cellars, economic potential. Looking at the site as a landscape in the sense of lifeworld for dialogical interaction is therefore a useful way to understand Carlsberg's spaces. It is a reading in which human agency is not separate from that of plants, soil, water and topography. By emphasizing the interplay between

57 It is classified as a special water reservoir because of its aquifer, *Miljøvurdering og VVM for Carlsberg Valby*.

different actors, we can see that the relationship between people and the hill is not static. The water situation exemplifies how this influence goes both ways: the industrial complex adjusts to where there are water pockets and sometimes architectural ideals must be given up because the water turns out to act differently than the way people had thought. Likewise, the intention to turn the site into planar surfaces was only partly possible, as the hill 'popped up' even more dramatically on the slopes in between those surfaces. People who interpret and shape natural landscapes are not the only agents affecting the physicality of open spaces; people exist as part of an interactive process with, for example, cultural, hydrological and topographical forces.

Industrial processes have left distinct footprints and have dramatically changed the open spaces of Carlsberg. However, these social, technological and economic rationales did not lead to action in a vacuum. Rather, new solutions were developed in a dialogue between industrial logics and architectural ideas. The first cellars were not built solely as technological spaces but were integrated into an aesthetic programme that emerged from the aesthetic ideas of English gardeners. The tanker route turns out to be connected to a differentiated social network yet it has been largely overlooked in the redevelopment process. The tanker route became subject to attention only when the financial crisis inspired alternative strategies and a pragmatic short-term reuse of physical structures that were already there.

Looking at the way open spaces have been used in the past could suggest new uses and positive associations that current plans and heritage inventories cannot inspire. Some open spaces were sites of the vernacular practices; such as the naming of certain spaces that have been associated with particular value, and the practice of using the slopes as social hubs. One may ask how important it is to know such practices, since the people who named and used the spaces are mostly gone. However, the examples of everyday uses of the former brewery landscape shown here enabled us to see aspects of the spaces that are otherwise often rejected without further notice, and so to appreciate the green slopes as more than leftover spaces. Knowing such characteristics adds on to the field of opportunities far beyond canonical heritage surveys and universal ideas about the good city.

The strong coherence of Gehl's concept of city life with the framing of the redevelopment process and the winning entry demonstrates that architectural proposals, like other cultural utterances, are not isolated. Rather, design projects should be understood in relation to the setup to which they respond. The way a competition is organized, the words that are used in the brief, and the formal and informal collaborations involved in a redevelopment process all affect how a site is understood and ultimately dealt with. It is thus vital to foster a critical discussion of how a redevelopment process is constituted and to reflect on how and when the steps in a redevelopment process – heritage inventories, competitions, plans – steer the path for future acts.

4 Sub-terrain

One particular part of the Carlsberg breweries became a central theme in many proposals for the international ideas competition 'Carlsberg Vores By' in 2006-2007: the cellars, 11 kilometres of them, beneath the brewery.[1] In this chapter they will serve as a locus to study how layered accumulations of myths, language, and human perception all play a role in the ideas and decisions generated about the future of an industrial site.

During the redevelopment of Carlsberg, it seems to have been troublesome to know how to deal with the cellars, which are not as easily reused in an economically profitable way as many of the buildings. The competition brief therefore asked 'Can the Carlsberg cellars be used for new purposes in the future?'[2] During the urban development project, the public gradually became aware of Carlsberg's cellars. Since the middle of the 2000s, increasing public access to Carlsberg brought enthusiastic urban explorers, neighbours, artists and others to go on tours of the cellars organized by Carlsberg Ltd.

After the building project was put on hold in 2008, little was mentioned about the idea of making the brewery cellars accessible for reuse. Rather, a new structure of parking cellars and an underground water reservoir were built independently of the existing cellars, which are now closed off. This is striking, thinking of their significant role in the brewery processes and in the imaginary of the breweries. Carlsberg's subterranean architecture has been connected to different values, allowing them to be perceived, used and their future imagined in different ways.

When the spirit of Carlsberg resided in its cellars

'An act of genius.' With this strong endorsement, journalist Synne Rifbjerg praised the present urban plan for the Carlsberg site by Entasis in an article in 2008.[3] In her view, the plan's main quality is that it ensures a desired vivid

1 Information from Carlsberg LTD Properties. The cellars are partly 3-4 storeys and overlap in various ways. The number 7-11 km depends on the definition of what a cellar is and is hard to measure. The main arguments in this section have been presented in an earlier version as Riesto and Hauxner, 'Underliggende Grunde.'

2 Carlsberg Ltd Properties, 'Vores By Carlsberg,' 45.

3 Rifbjerg, 'Brygge og Bygge.'

urban life while at the same time revealing the 'spirit of the place'.[4] Rifbjerg argues that the urban plan is able to keep the spirit of the place 'intact' by relating closely to the plan of the cellars.

This appreciative critique refers to the way that Entasis develops a new street/square pattern for Carlsberg. The masterplan partly adjusts to existing roads and buildings. Most important for the proposed urban form, however, as the architects explain in their proposal, is what lies beneath it: Carlsberg's cellar network. The plan drawing of the cellars was repeated on every page of Entasis' competition entry as a recognizable logo for the proposal (4.1). The proposal presents an easily communicated concept: wherever there is a cellar today, a square will be situated in the future with the exact same plan; wherever there is a cellar corridor there will be a street that mirrors the cellar beneath it (4.2, 4.3).

Image 4.1 A plan of Carlsberg's existing brewery cellars became the logo and key creative reference for Entasis' winning competition proposal 'Vores Rum' (Our Spaces) in 2007.

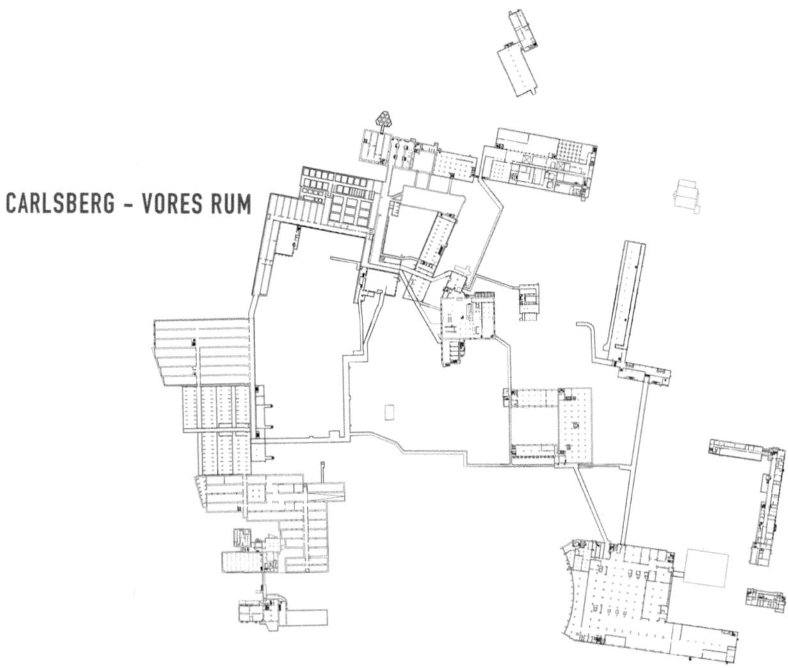

Picture of the front page of the proposal

4 Ibid.

Image 4.2 Analysis from Entastis' competition proposal from 2007. Black depicts the cellar plan projected onto terrain level. Grey shows the existing buildings to be preserved.

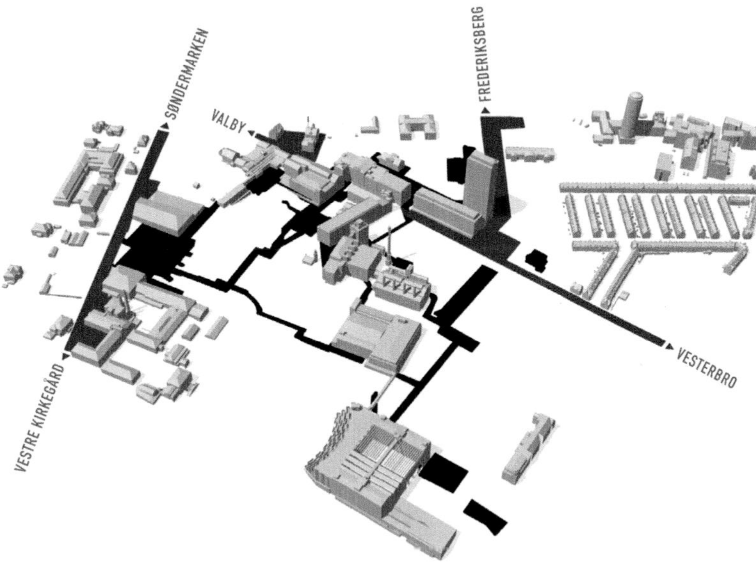

Image 4.3 Entasis has projected the plan of the cellars up to street level and let it form the new street/square pattern and the relationship between mass and void. Dark grey depicts 3-5 storey buildings and black highrises.

Image 4.4 Narrow new street built according to Entasis' plan with three-to-five-storey housing leading up to a square. Construction work in 2016.

While relating to only one urban form principle, derived from the cellars, however, the proposal adjusts only nominally to the existing building/open space relationship on ground level. For example the plan does not incorporate the large open spaces from the late twentieth century that were part of the driving route. The result of Entastis' overall concept is a paradoxical masterplan that appears as if it was not drawn by architects. Instead, the new streets and spaces seem to be the result of long-term processes, like an old city that has grown according to the changing needs of building, adjusting roads and more. Nevertheless this plan prescribes the footprint of the new buildings and inserts a new urban form with narrow streets and enclosed squares all over the site (4.4). As the analysis of Carlsberg Square showed, the Entasis architects do not engage explicitly with the multi-terraced topography that the cellars have cast off by being underground – the plan of the cellars themselves is what interests the architects. While the journalist referred to the cellars as Carlsberg's true essence or spirit, the architects, in fact, relied on a different understanding of the subterranean architecture.

In the winning competition proposal, Entasis presented some ideas as to how the cellars could be used for activities like sport, wellness and art (4.5). For example, if the roof of New Carlsberg's storage cellars is removed, the spaces could be filled with water to make a public bath and Old Carlsberg's cellars become a spa, as can be seen on the cover image of this book. When the plan was prepared as an official development plan in 2008, Entasis, Copenhagen Municipality, Carlsberg and other actors negotiated possible alterations to the plan. The ideas about reusing the cellars were discarded in this process. From the perspective of Entasis' client, Carlsberg Ltd, the cellars were exciting spaces which could be used to host events and to brand the new urban district as something very special, but they possessed no direct potential for reuse and income generation. The idea of reusing them was quickly discarded in the negotiation. What was not compromised, however, was the layout of the new urban plan, which was derived from the cellar plan: every new building, for example, had to align to the niches and corners of Entasis' masterplan, derived from the cellar plan. Rather than perceiving the cellars as three-dimensional spaces and shapers of Carlsberg's terrain, Entasis is concerned with the built form of the future city district. This also has a practical and economic aspect that fits well with Carlsberg Ltd' ambition to ensure a profitable urban development; cellar decks can serve as floors for the new squares, while the parts of the site with no cellars underneath can more easily become the ground for the construction of tall buildings.

After winning this large competition, Entasis founding partner Christian Cold presented the plan in discussion panels, seminars and public meetings. To describe the desired atmosphere, Cold showed pictures from urban spaces in the city centre of Copenhagen and the Italian towns Siena and St Gimignano. The fact that this preferred urban form could be linked to the plan of the industrial cellars in this competition proposal appears to be only a fortunate coincidence that happens to benefit the creative process and rhetoric of the project.

Image 4.5 Ideas of how the cellars could be reused for new purposes, as depicted by Entasis architects in their winning competition entry.

Entasis' engagement with the cellar plan can be further understood by two concepts promulgated by the French philosopher Sébastien Marot. Marot explores the relationship between people, memory and physical surroundings in his research, which is closely connected to contemporary design practice. His starting point is a critique of design strategies that do not take the specific site into close consideration but works '"from the inside outwards", from program to site, from the city to the territory'.[5] Marot argues that designers should engage thoroughly with each specific situation and never let abstract programmatic ideas overshadow the existing site. To describe an approach that starts with the concrete location and not the general brief, he introduces the term 'sub-urbanism'. Super- and sub-urbanism are not absolutes, as Marot repeatedly emphasizes. Every design project obviously deals with both programme and site, with the general and the specific but Marot's concepts become useful to understand how design occurs on a scale between the two extremes.

Entasis' plan is not the result of a thorough engagement with Carlsberg's cellars. Rather, it depends on a general idea about how to design for vivid and populated urban spaces, a super-urban approach. Entasis appear more interested in fighting a battle against the modern city than engaging thoroughly with the specificities of the Carlsberg site. The cellar plan here serves as a conceptual source of inspiration and as a rhetorical figure to justify an idea about 'classical' enclosed squares, narrow streets and a dense urban fabric. Interestingly, these ideas, which are rooted in the critique of modern architecture and urbanism, here correspond well with that of Carlsberg Ltd, which wants to build densely and optimize the economic property value. Hence, what at first glance – and through the lens of the press – appears to be a plan particular to this site is rather born out of generic ideas. Entasis is thus not unearthing the 'spirit of the place' in the underground – itself a dream that cannot come true – but presenting an idea about the 'good city' and helping the owner optimize the property value.

Chapters in the biography of Carlsberg's cellars

There are many ways of reconstituting aspects of Carlsberg's underground that adhere to it more specifically – both its history and the experience of them today. Alternative readings of the cellars point towards other values and possibilities for action. In the following, I sketch the technological

5 Marot, 'Sub-urbanism/super-urbanism', 21.

history of this subterranean network and the cultural practices and myths that have evolved along with it.

Technological histories of Carlsberg's cellars

Carlsberg's cellar system evolved in a long process of shifting production and distribution rationales (4.6).[6] The first cellars, begun in 1847, were isolated vaulted fermenting and storage facilities necessary to produce the dark beer that became Carlsberg's success (4.7). The cellars were expanded several times in the late nineteenth century. Beginning in 1876, a subterranean corridor connected the cellars to the Villa Faxehus for Erhard Kogsbølle, manager at Carlsberg (4.6, 1876). The MB Brewery was placed orthogonally to the Old Carlsberg site on a former sand pit, while the New Carlsberg cellars were dug where the terrain slopes significantly, as can be seen on the diagram of Carlsberg's cellars in the year 1900 (4.6). Those new brewery cellars are much larger than the ones of the Old Carlsberg complex.

When Old Carlsberg and New Carlsberg merged in 1907, a connecting subterranean corridor was excavated between the two cellar systems. Production lines were now connected into a new shared mega-system. No new fermenting and storage cellars were excavated because beer could now be fermented above ground, due to the cooling machines that had been introduced to Carlsberg in the 1870s.

In the twentieth century Carlsberg's cellars were expanded numerous times towards the east. New cellars were built to serve new functions, developing into a complex that was considered the fastest and most rational transportation of various types of ingredients and manufacturing components (4.6, 4.8). The cellars also conveyed electricity from the central power station built in the early 1920s.

Around 1910 beer was pumped to the bottling plants in subterranean pipes for the first time, which radically reduced the need for manpower and mechanical power to transport the liquid.[7] The transportation factor in the increasingly integrated system of production occurred underground, and the cellars facilitated the movement of wort, beer, water, electricity, steam, carbonic acid and ammonia for cooling (4.9).

6 The following analysis is based on Glamann, *Øl og Marmor*; Glamann, *Bryggeren*; and Jørgensen, 'Carlsbergs bebyggelseshistorie.' The maps were first sketched during personal communication with Ulla Nymand and Niels Nielsen, who generously shared their knowledge about Carlsberg's cellars.

7 Jørgensen, 'Carlsbergs bebyggelseshistorie,' 65-66.

Image 4.6 Expansion Carlsberg's cellars.

Carlsberg's cellars have been constituting for the development of the brewery site. They were changed and rearranged several times along with shifting technologies for brewing, lagering and distributing beer. Red colour shows what is added at every building phase shown here. With the ongoing redevelopment, new parking cellars are planned beneath some of the large buidlings. The existing cellars are for the most part inaccessible.

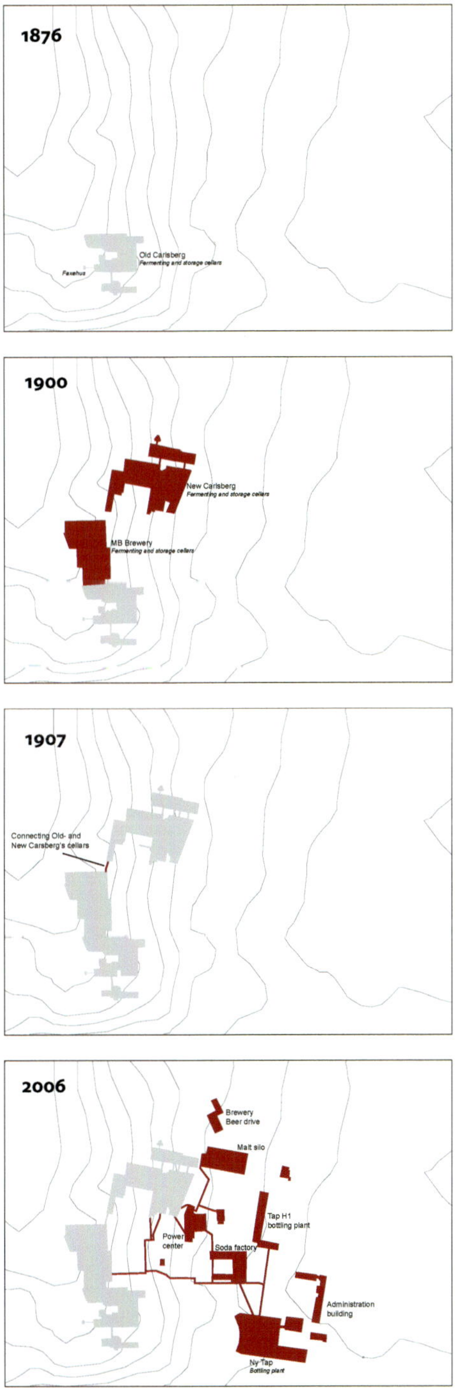

Image 4.7 Cellar from the first Carlsberg brewery.

Image 4.8 Cellar connecting corridor from the twentieth century.

Image 4.9 During the twentieth century, the cellars were a central part of the industrial process complex and transported ingredients and manufacturing components used at different phases of production.

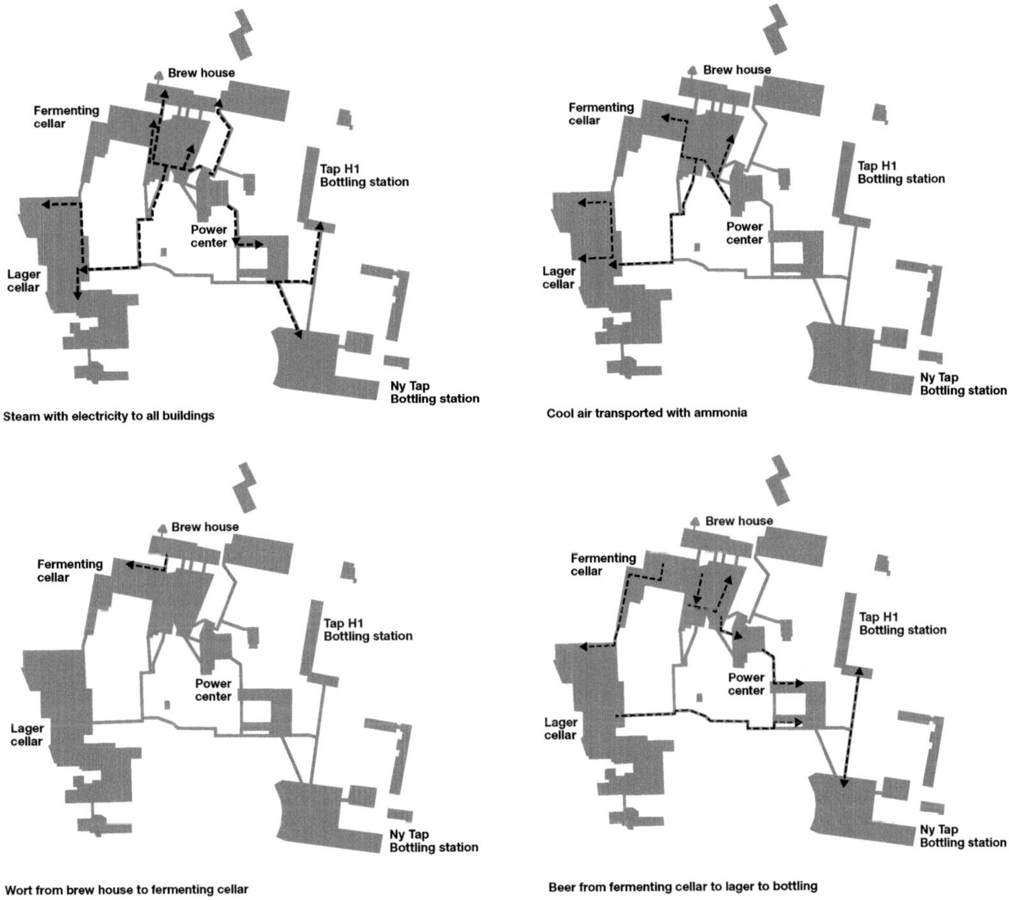

Much later, in the 2000s, these invisible flows fascinated contemporary artist Jacob Kirkegaard, who recorded the sound of substances transported through the multiple cellar pipes during the last years of production. He launched a composition of these sounds of the brewery in a New Carlsberg cellar in 2011, replayed in 2014, as a paraphrase of an audible trace of activities in the breweries. Kirkegaard was not the first visitor to be fascinated by these cellars. Since the mid-nineteenth century foreign brewers and scientists visited Carlsberg on guided tours, appreciating the cellars' technological capacities.[8]

Myths and uncanny cellars

From the beginning the cellars have been experienced as places of discomfort. In 1881, the novelist Herman Bang expressed this very well.[9] Bang was concerned with the danger for workers who spent long hours in the fermenting underworld beneath the breweries. During the fermenting process, he knew, carbon dioxide poisoning could occur and the cellar workers were exposed to a fatal risk.[10] Yet, despite (or because of) this uncanniness, Bang is emotionally drawn to the cellars. In this report, which is full of pathos, Bang shares the excitement of experiencing an intense atmosphere. He repeatedly compares the subterranean facilities with caves in ways that make the visit to the brewery sound like a sublime experience of nature. Bang visited the cellars with one of the breweries' guides on a day when workers were cleaning the cellars by throwing water on the walls and stairs. He tells us:

> With every turn one feels one is standing before a mouth of darkness. And it becomes colder and colder. 'It is like a chasm in the mountains,' one says down in the darkness, where a flicker of light is seen around the next corner. You must hold tight as the steps become wetter.[11]

This ambiguity between fascination and fear also characterizes employees' perceptions of the cellars. One horrifying event, which became a story often told by workers, left one of the cellars with the colloquial name 'Corridor of Death.' The story goes that when Old Carlsberg and New Carlsberg merged in 1907, peace arrived after decades of competition and a sometimes icy relationship between the two companies.[12] The decision to merge was made after the reconciliation of the members of management of the two

8 See, e.g., Zanker von Meyer, *Die Bauten von J.C. Jacobsen und Carl Jacobsen*, 37.

9 Bang, 'Et Besøg paa Carlsberg.'

10 Ibid.

11 Ibid., translated by Richard Hare. Original quotation: 'Ved hver Dreining synes man, man staar foran et Gab af Mørke. Og der bliver efterhaanden meget koldt. 'Det er som en Bjergskakt,' siger man ned i Mørket, hvor man ved Dreiningen ser ens Lys flakke. Man holder sig saa fast, fordi det begynder at blive fugtigt paa Trindene.'

12 Nymand, personal communication, 31 October 2008. When I visited the site in 2007-2008, I also had many conversations with workers who confirmed – and retold – this narrative.

Image 4.10 Cellar situation depicted by an urban explorer in 2008. The shop dummy was a running joke amongst brewery workers, who would hide it at different places in the cellars to scare their colleagues.

breweries. The workers, however, were not yet used to the unification. When the corridor between the cellars of the two breweries was being dug, the workers got into a fight, resulting in substantial injuries and fatal consequences, as the name suggests.

In the twentieth century ammonia contamination has been the main risk of working in the cellars. As a safety precaution, only a very few workers have had access to the cellars and, as a result, few have known how to navigate their maze-like structure. In the last years of production this small cohort of workers had an ongoing practical joke down in the cellars. To enhance the thrill of their colleagues' experience and the potential danger found in the cellars, they placed various parts of a shop dummy in different locations in the complex (4.10).[13] What would a visitor find behind the next corner in this spooky underworld?

13 Nymand, interview, 7 July 2011.

The cellars – A fascinating locus for imagination

The contemporary negotiation about how Carlsberg should develop into a 'good city' cannot be separated from what has been spoken, thought and felt about these cellars. Carlsberg's massive subterranean architecture is not simply functional but it is also a rich source for stimulating human imagination, triggering our fear and desires. The celebration of the urban plan's relationship to the cellar plan can be understood within the backdrop of a broader cultural imaginary.

Associating the underground with value has a strong tradition in Western culture. For example, in Marot's writings, for his concept of a desired design strategy, he uses the prefix 'sub', which means 'below'. The 'sub-urb' has been thought of as less important than the metropolis in the traditional concentric idea of the urban. Marot elevates the architecture in the periphery and associates it with value. But the prefix also refers quite literally to the sub-terrain. Marot writes that 'sub' means 'the ground on which the city is founded'.[14] He often refers to the physical underground with such striking expressions as 'deepening the territories',[15] 'digging below the surface'[16] and 'grafting'.[17] These expressions are not particular to Marot's writings but connected to a cultural tradition in which the sub-terrain is considered closer to the truth than what stays on the surface. The English distinction between 'superficial' and 'deep' can also be found in many Germanic languages.

In her book *Notes on the Underground*, historian Rosalind Williams investigates the special role the sub-terrain plays in Western culture.[18] She explores how a long-standing cultural tradition encourages us to distinguish vertically between our familiar surroundings and the world below, an underground that is associated with something substantial and essential, as an 'other' space. Williams argues that this distinction relates back to the Enlightenment, when, she argues, the idea that digging came to have the connotation of revelation. At the same time practitioners in the emerging disciplines of geology and archaeology started to carry out empirical excavations. Researchers began digging for strata and other evidence that countered long-standing beliefs about the origin of the earth.

14 Marot, 'The Reclaiming of Sites,' 55.
15 Marot, 'Sub-urbanism/super-urbanism,' 21.
16 Marot, 'The Reclaiming of Sites,' 56.
17 Ibid.
18 Williams, *Notes on the Underground*.

Image 4.11 A Carlsberg cellar situation as depicted by an urban explorer in 2008, when most of production had moved and the cellars had lost their original function.

Williams concludes that the association of the underground with the truth has become so ubiquitous in our culture that it often goes unnoticed: 'Since the nineteenth century [...] excavation has served as a dominant metaphor for truth-seeking. The assumptions that truth is found by digging, and that the deeper we go the closer we come to absolute truth, have become part of the intellectual air we breathe.'[19]

The journalist who celebrated Carlsberg's cellars as the spirit of the place is a carrier of this imaginary. She conceives of the cellars as a particularly deep and substantial aspect of Carlsberg – carrying its spirit – and hence relies on the clusters of metaphors of the underground that are inherent in Western culture. The cellars play an important role in the perception of Carlsberg not only because they are crucial actors in the brewery's fermenting and distribution but also because they are connected to certain

19 Ibid., 49

cultural ideas. The cellars are the result of an additive-building process that facilitated yeasting and fermenting and, later, transportation in a large industrial complex. However, the cellars are also a cultural and practiced space, and our access to them is coloured by a broader cultural imaginary. Entasis' plan is the result of a creative engagement with the underground in which the overall concept is modified to fit with the cellar plan.

This may seem like a process that is quite different from digging your way down to the truth about a place through scientific excavations but Williams points out that empirical science and excavations have not replaced myths and narratives about the underground with the truth. Instead, the thinking connected to excavations has altered previous myths and contributed new narratives. Williams argues that 'scientific inquiry retains an aura of the mythological, since the heroic quest for scientific truth has the pattern of a descent into the underworld'.[20] This scientific activity might not be so different from Entasis' story of the development and decline of the classical city and reuse of the engineers' cellar plan as a verified document on whose base Carlsberg can reoccur as a 'good city'.

Entasis used the cellar plan to legitimize an urban form of the lost but longed-for, premodern city. The firm's association of a complex subterranean production structure with the premodern city reveals that any engagement with an industrial site is a creative process. It is thus extremely reductive to propose that the spirit of the place is a given fact, as suggested in the press. Entasis' masterplan does not reveal the spirit of Carlsberg, and neither does any other design proposal or survey. Instead, it is more relevant to discuss Entasis' plan from another point of view: to question if the classical European city is a universally applicable urban ideal. This is the main assumption underlying Entasis' masterplan, for which the cellar structure is used as part of the argumentation. We do not have access to the 'pure' Carlsberg cellars, or any other spatial structure. Instead, the cellars have already been interpreted, and reinterpreted, in ways that influence our current perceptions and choices as to dealing with their future. The underground – including Carlsberg's cellars – remains a creative locus for the human imagination and interpretation.

This book seeks to contribute to building an understanding of former industrial sites as dynamic, multidimensional participants in larger urban landscapes. My initial aim was to study transformation rather than stasis and to investigate how the interaction between people, practices and environmental processes works in a particular context. Digging into Carlsberg's

20 Ibid.

cellars exposed how landscapes are not only characterized by change, but also by old cultural traditions that work over a longue durée. So while contemporary writings about urban landscapes often highlight change, Carlsberg's cellars is a reminder of the long-standing cultural frameworks that, often quite secretly, operate in and potentially colour our decisions during urban redevelopment. Further, while Chapters 2 and 3 explored how Carlsberg's landscape from a relational perspective, studying different relationships between people's practices, cultural and environmental processes in unbounded geographical constellations, this chapter has explored another kind of relationship – that between underlying cultural imaginaries and our thinking and doing in the present.

Conclusion: Biography of industrial open spaces

The last decade's redevelopment planning of the Carlsberg breweries shows how different perspectives from different fields came together in a strong alliance: The authorized heritage discourse directed its attention towards a few *old, grand and beautiful objects* made for those in power.[1] This selective view on heritage formed a strong alliance with the economic agenda of Carlsberg's urban development project; it left a lot of space to erect a dense tissue of new buildings that could increase the property value. This alliance became reinforced when Entasis architects delivered a powerful story behind the urban plan – building dense was now presented as the most sustainable solution; The theory was that dense districts reduce the need for car traffic because people live close to work and because they can walk, bicycle or use public transportation. This stated urban ideal, however, did not restrict the urban project from including large new parking cellars, which will support car transportation. Further, the architects saw the dense city as especially suited to create what they saw as vivid urban life. The fit between the density ideal in urbanism and profit-oriented property development is not restricted to Carlsberg, but a characteristic of much contemporary urbanism, as for example, it is discussed by Thomas Sieverts.[2] Yet, the case of Carlsberg shows particularly the extent to which the authorized way of doing cultural heritage contributes to this alliance by concentrating on a few built objects in the urban landscape, but not offering any basic resistance to the economically driven urban development.

The dense city ideal was contested several times during the development. Notably by the competition entry 'Time Will Tell,' which resembles a point made by geographer Tim Edensor, who has written about the role that abundant industrial sites play in contemporary society. Edensor perceives the derelict production site as a counterpart to the contemporary city, where every square metre is commodified and determined for certain functions, owners and mechanisms of exclusion.[3] In a situation where economic

1 Some of the arguments in the following section were presented in an earlier version in Riesto, 'The Old, the Pure, and the Quirky: Contested Heritage Values in the Urban Redevelopment of the Carlsberg Breweries in Copenhagen.'

2 Sieverts, *Zwischenstadt*, 44.

3 Edensor, *Industrial Ruins*.

forces are very decisive in city making, leftover industrial sites can prove a special value in not being programmed and steered by fiscal agendas. The entry 'Time Will Tell' seems to work from such a premise and thereby attempts to create a district that is outside of the fundamental conditions of capitalization and control of space.

While everyone quickly forgot 'Time Will Tell' after the competition, there was a period when this competition proposal was perhaps the closest to being true; after the financial crisis in 2008 paused all planned construction projects and Carlsberg Ltd initiated a phase of new users and uses. Urban explorers, artists, gallerists, skaters, a go-kart company, fairs, parties, and a supermarket entered the area (5.1). In this phase the Carlsberg brewery site became increasingly well-known in the public and a popular cultural hub. Yet in 2015, when new investors arrived and the development that had been planned prior to the financial crisis proceeded as if (almost) nothing had happened, the temporary rental contracts of the culture producers were stopped and many of the buildings that they had used were demolished. As the construction of new buildings begun in 2015, all Carlsberg's temporary urban space installations were demolished and most of the then popular urban spaces made inaccessible because of construction work.

It was in this period that Carlsberg Ltd begun to be criticized for kicking out a unique possibility for a vivid district – the end goal after all.[4] Ironically, critics mourned exactly those aspects of Carlsberg that hardly anyone had paid attention to in the official heritage assessments and architectural proposals for Carlsberg: the apparently chaotic 'terrain vague' feel, the programmatic openness of large twentieth-century industrial distribution buildings and the vast open spaces shaped by the industrial production in the 1950s-1990s. There is a striking gap between their perception of twentieth-century Carlsberg as 'edgy'[5] and characterized by an 'aesthetic of decline, 'terrain vague' and industrial charm[6]' and the established valuation that had laid the basis for official heritage decisions and the urban plan. Obviously, many of the gallerists and event makers that stayed at Carlsberg are first movers, who were ready to leave for the next place to discover, anyway. This difference exposes the fact that the processes of heritage making in an urban landscape involve dissonances and conflict and that they are tied to questions about the power of different actors and

4 See e.g. Holst and Jacobsen. 'Visionen der forsvandt.'

5 Urban Consultant Mads Byder from the bureau Urban Help quoted in Fejerskov, 'Carlsberg Byen lukker populære kulturtilbud på stribe.'

6 Rosenvinge, 'Jeg drikker aldrig mere en Carlsberg.'

Image 5.1 As new users became familiar with the Carlsberg site around 2010, they expressed admiration for the 'edge' and 'quirkiness' of the area. Surprising juxtapositions between old and new elements created an atmosphere that in their view made the site special. This outdoor climbing facility was built into a row of trees that separates a statue of the brewery director Carl Jacobsen from a production area in the back. It became a popular place to visit and play in a special atmosphere, and is now removed.

groups in society.[7] The point here is not that the first-movers at Carlsberg, many of whom were from the art and culture scene, should be upheld as a new authority and their appreciation of 'industrial flair' become a new norm. Rather, their new take on Carlsberg's values exposes the dynamics and contested nature of heritage, in which a broad range of actors represent different, and often changing values.[8]

To acknowledge the dynamic character of heritage values, the way that redevelopment projects are structured must be rethought. When Carlsberg's

7 Tunbridge and Ashworth, *Dissonant Heritage.*
8 Riesto and Tietjen, 'Doing Heritage Together.'

production was relocated there was a sudden shift in users and thereby the meaning-making of the area was abruptly changed. yet, the planning process, which relied on expert views, made no openings to adapt to their perspectives and learnings from the new user groups. Neighbours to the breweries and potential new users had previously been asked to react to the plans for the district, following the practice of public hearings, which is embedded in the Danish planning legislation. But it was only after the realization had begun that new users had got to know the formerly closed-off area and began experiencing how its buildings and spaces could be reused and reworked, that a public debate emerged. Also, it was only in 2015, when people were able to see the physical effects of the plan – industrial buildings being torn down, old trees removed and high-rises being erected – that they began to criticize the plans. Finding more adaptive ways of planning is one of the key challenges in redeveloping industrial urban landscapes.

Further, the thinking behind architecture preservation that came to impact on Carlsberg's transformation has not been explicitly stated, but can be studied up close on those buildings that are kept in the new district. Buildings that are listed – large monumental buildings like the machine centre, and the kettle house – have been peeled so that the messy industrial equipment (pipes, ladders and equipment that physically connected them to other buildings) that once characterized their exterior is now gone (5.2). What is left is more or less solitary monuments with little or no reminiscence of the mess, danger and dirty work that were also part of this brewery. Also, the multiple building additions and alterations – loggias, small guard houses, wooden sheds – that testify the many uses and temporal layers of this site, have largely been removed.[9] This cleaning process was driven by an urge to emphasize historical qualities of the buildings, but it paradoxically also creates an architecture that never was in the breweries before.[10]

A broader perception of what can potentially be valuable contributes to making our cities more aesthetically diverse and compatible with the city's social and cultural plurality and dynamics. In the encounter with an existing urban landscape, planners, designers and heritage managers' ideas about what is beautiful and valuable can be modified and new solutions can be developed. There are no obvious answers to what should be demolished,

9 This is related to a discussion initiated by art historian Max Dovrak about 'parasite buildings' versus monumental buildings in architectural preservation. His viewpoints are still highly relevant, as shown in, e.g., Lægring Nielsen, 'SAVE som politisk og æstetisk praksis.' In Tietjen, Riesto, Skov, *Forankring i Forandring*, 162.

10 For a further examination of the idea of the pure in architecture preservation, see, e.g., Raskin, 'Jorge Otero-Pailos.'

Image 5.2 The old, monumental buildings at Carlsberg that will remain standing between the new buildings are usually cleaned and polished. Here, pipes and other structures that had been attached to many the walls on the machine centre are removed. Thereby some of the characteristics of this urban area disappear, both in relation to documenting a part of the social history and to creating a sense of wonder in the future urban quarter.

reused, preserved and how, but there is a need for lively debate about values and about how different voices can make themselves heard in the change of urban landscapes.

Understanding industrial open spaces in the context of urban redevelopment

The case study of the transformation of the Carlsberg breweries shows that industrial open space is still a relatively ill-defined category in heritage management and spatial design and planning; the concepts applied are often implied rather than explicitly expressed, and many aspects of the open spaces are not grasped by the methods and perspectives applied. In the following, I will outline some theoretical and methodological insights in terms of understanding industrial open space that emerge from the empirical study of Carlsberg and which can hopefully contribute to a much-needed discussion.

The first point concerns the necessity of broadening the concept of open spaces that are is applied to industrial redevelopment sites. Throughout the redevelopment of Carlsberg – an innovative heritage-in-planning process with great focus on urban space – the most decisive actors quite seamlessly reduced Carlsberg's existing open space values to a few enclosed, axial squares and thus automatically ignored or rejected the spatial structure on most of the brewery. Such a canonized preference for pre-industrial European town squares does not enable us to grasp industrial sites in general, and especially those built after World War II, on their own terms. Research can play a vital role in exploring industrial open spaces from other perspectives, such as their vast, large surfaces, their role in flows and routes of industrial processes, and their role as lived spacess with histories and discourses in their own right. Grasping such characteristics requires us to develop specific methods of survey that are adjusted to the specific urban landscape under transformation and to use spatial concepts that can grasp the characteristics of the given spatial situation.

Futher, the concept of open spaces can be broadened by abandoning the view of them as fixed form, and engaging in their dynamics. Open spaces as presented here are continuously emerging as an interplay of different spheres. Lefebvre's triadic spatial concept allowed for a rich understanding of industrial open space as spaces with thick life stories and characteristics in the dynamic interrelationship of materiality, practices and discourse. This interaction did not constrain itself to any scale, but characterized both the participation of industrial open spaces in larger road and distribution networks, and in cultural discussions with a much further reach than each space itself. Therefore, studying relations is not a new universal truth or norm, but a cultural activity which can, however, substantiate how we participate in the ongoing transformation of sites. In short, to intervene in tomorrow's processes and relationships that a site participates in, we benefit from knowing those that are operative today and how they became.

The third point is that industrial open spaces involve human dissonances, such as the Carlsberg workers' strikes for better working conditions in the 1970s in opposition to the company management – an event often not told about in the corporate storytelling. It may not come as a surprise that the marketing department of Carlsberg Ltd used such a simple storyline that focused on heroic histories to an extend that it blended out the social and political histories of Carlsberg. But that it was backed up by almost all heritage professionals, planners, urban designers, landscape architects and architects that worked at the brewery is striking.

The social orderings and conflicts in Carlsberg's open spaces are not only played out *in* Carlsberg's open spaces, but at some points also *with* the open spaces so that the very materiality participates in the social ordering. This is what happened when the bottlery station Tap 1 was extended with a semi-circular façade that separated the workers from the large landscape garden, then reserved for dwellers in the honorary residence. The façade that architect Svenn Eske Kristensen designed for this extension did its job; it prevented bottlery workers from looking out and at the same time seemed to be a green wall when seen from the garden. This wall was later loaded with different meanings, and got nicknamed 'The Hanging Gardens'. This interplay between materiality and social practice can be grasped with a biographical perspective, beyond distinctions between tangible and intangible heritage or discussions that focus on architecture as artistic expression.

Three ways of understanding change with Carlsberg

Studying Carlsberg enabled a close look at what it means to conceive of a landscape in a biographical perspective – namely, how to conceptualize changes (and persistence) over time. What is it that constitutes the temporalities of landscapes? Recent research has called for qualification of what is often (for a lack of better phrase) called the 'temporal layeredness' of landscapes.[11] Thinking of landscapes through metaphors such as layered

11 Renes, 'Layered Landscapes.'

and palimpsest can be a productive start to describe an area like Carlsberg, which, has not evolved in a linear process, but, as we saw, has gone through a various long-term and short-term reconfiguring processes. But how can we untangle the processes that produce landscapes in infinite processes? I identify three aspects of landscape change that emerged from studying Carlsberg from a biographical perspective: First, that Contesting ideas shape landscapes; second, that humans shape landscapes together with cohabiting actors and third; that inherited cultural imaginary is a force in landscape formation[12]

Contesting ideas shape landscape

Landscape narratives are not just abstract ideas that people attach to landscape. Rather, the way that people conceive of landscapes are forceful and influence how we think and act together in and with landscape. For instance, looking at the contemporary redevelopment at Carlsberg: different professionals applied different preconceptions that steered how they perceived the Carlsberg area as it is now and how they proposed it to change in the future. Vogt Landscape Architects' ideas of establishing a new role for nature in the city contested Entasis' preference for the classical urban square which they believe to be a vehicle to a vibrant city life. The discussion about how to define Carlsberg as a site can be understood as a battle between different perspectives. The perception of the entire Carlsberg site as being equal to the brewery's locus of control strongly influenced the inventory of listed buildings. In the heritage assessment of the site, alternative and relational concepts of the site were proposed of Carlsberg as an industrial site and by designers who put their attention on reinforcing its relationship with the city in different ways.

Such processes can be understood as different narratives and values that contest each other. Analysing a site according to this model implies focusing on how different actors with different sets of values negotiate possible ways to understand an urban landscape and to intervene in it. The stakeholders in a redevelopment process rely on different disciplinary mind-sets and specific 'habits of mind'.[13] Their worldviews can be defined as belonging to disciplinary lineages, such as the heritage experts, who saw Carlsberg as a

12 Although emerging from the empirical study of Carlsberg with the lenses of landscape biography and Lefebvre's concept of space, the three points that I make here are also similar to what science and technology scholar Anique Hommels did in the book *Unbuilding Cities*.
13 Burns and Kahn, *Site Matters*, 314.

cultural-historical site, was at odds with those who confined themselves to what was within the present-day property line. The values that we apply to a site do not, however, necessarily match disciplinary borders. Professionals from the same field can apply different lenses, as the differing readings of the Carlsberg site posed by various designers exemplify. A landscape, then, changes when different actors with different worldviews conflict. The actors that are most dominant at a given point in time become the ones that change the landscape, for example, by deciding what is reused, demolished, listed or altered, and how.

As the study of Carlsberg shows, when an encounter between different sets of values occurs, the outcome is not necessarily always a battle. Rather, sometimes actors with different frameworks can form unexpected alliances. An example of this is the negotiation of the future of what was named the Carlsberg Square in 2008-2009. All the professional actors with the power to make decisions agreed on certain aspects. The corporate aim for a profitable urban project coincided with the desire of Entasis and the municipal planners to ensure the project entailed classical urban activities and a vibrant city life. This vision corresponded well with the goal of representatives from the Heritage Agency of Denmark, who were occupied with listing buildings. According to their framework the small boiler houses could only be worthy of listing if their material state could be defined as authentic, which the boiler houses could not because their interior had been altered in the 1960s. Although these professionals saw the boiler houses from very different perspectives, their conclusion was the same: The boiler houses had no value worth the effort to take any concrete action to preserve them. These professionals relied on radically different frameworks, each with their own pedigree and rationales but, for a while, their interests overlapped.

This shows important aspects of how open spaces are conceived. It opens up for the possibility of a refined understanding of how open spaces are used, perceived and altered by different actors. As the landscape biography of Carlsberg shows, not only textual sources give us insight into such discourses. They are also found in heritage surveys and design entries for the ideas competition and the subsequent urban plans for the site. Studying design proposals implies investigating many aspects, including how they are framed by the brief, mediation by the client, the rhetoric and critique of the project. The strength of understanding change as the interaction of different values and conceptions is that it allows us to critically examine underlying assumptions and when relevant think places anew. This way of perceiving change can deepen the contemporary debate in a redevelopment process, and is especially important because heritage and design practice sometimes rely on rationales that appear to outsiders as being tacitly accepted. Positions that used to be presented as self-evident can then be articulated. What cannot be covered with this thinking about landscape change as competing actors with different sets of values, is those actors that are not human.

Humans shape landscapes together with cohabiting actors

Not only people, but also other aspects of Carlsberg's landscape also participated actively in the production of space. To recapture that Kolen and Renes wrote about landscape biography: 'Animals, other organisms, maybe even "animated" objects and places (in specific cultural perceptions) or human-made technologies – all belong to the group of cohabiting actors that together make the landscape into a life world.'[14] Who or what can such actors be?

In the case of Carlsberg, substances like water, yeast and soil have been strong forces in the formation of the brewery's open spaces, in interaction with people's shifting practices and conceptions and with other materialities, such as the geological fault. In other words, the people who shaped Carlsberg's open spaces have not acted in isolation. Valby Hill itself, and its hydrological and geomorphological underground, are key to understanding how this site came about in its present state and may well turn out to be decisive factors for its future, for instance, concerning storm water events or heated temperatures caused by the climatic crisis. While the industrial brewers at Carlsberg in the 19th century attempted to control the water, the water pockets in turn forced them to modify their plans. For example, the desired symmetry and the separation of pleasure and work in J.C. Jacobsen's garden could not be maintained because of the need for a new well in the garden. The hydrological situation, therefore, is not a neutral, passive system or process, but an active force that has affected Carlsberg's physical development. J.C. Jacobsen's garden, therefore is not only a designed landscape that resembles aesthetic ideals of the mid-19th century, but testifies the cohabitation of human actors and courses of action that are not human.

Understanding change in this way can shed light on how various courses of action – brewers, water, wells – mutually change each other and form assemblages in the city. An example of such an assemblage is the interplay of factors involved in reaching the moment in history when one part of Valby

14 Kolen and Renes, 'Landscape Biographies,' 32.

Hill came to be the first Carlsberg brewery: the topography of the hill, the yeast's sensitivity to high temperatures, the brewer's idea of building cellars, the attractive cultural environment on Valby Hill, the new railway line that was under construction, the rich groundwater supply on this moraine, the favourable beer market in Copenhagen, and J.C. Jacobsen's ambition to brew bottom-fermented beer.

This assemblage of people and other organisms and substances explains how fallow fields became radically transformed into the first Carlsberg complex with terraces, production buildings and a landscape garden in the mid-1800s. Such assemblages are not necessarily local. The Carlsberg brewery's increasingly international trade and Copenhagen's growing property market led to the emptying of the site of production and reuse for cultural functions that slowly make their marks in the open spaces. The regional roads, the railway and the international distribution network also changed the significance and thereby the use of the site.

To acknowledge the interplay between people and other cohabiting actors adds an important dimension to landscape biographies. Landscape processes are key to deepening the understanding of industrial open space. But applying such a perspective in practice requires rethinking established practices of creating heritage inventories, which are still too often confined to static conceptions of space and binary thinking about nature conservation and cultural heritage. Grasping industrial open space in a landscape perspective also requires alternatives to abstract architectural approaches that apply a certain urban form to industrial sites without dealing closely with what is particular to each site. Landscape processes should not, however, be thought of as essences or as an apolitical zone free of economic interest and cultural interpretation, but should be integrated into a critical heritage and urban debate.

By conceptualizing change in this way, we can study the interrelationship between soil, trees, human culture, technology, water and other elements over time and thus illuminate how an industrial landscape is more than the product of people. Yet, the example of Carlsberg also shows that there are long-lived cultural traditions that also play a special role in forming landscapes; previous interpretations influence the future development, as will be the focus of the next section.

Inherited cultural imaginary as a force in landscape formation

Many an idea and sensorial thrill have been projected on the underground component of the Carlsberg site, be it the search for the 'spirit of the place' or in the workers' practical jokes as a way to deal with the uncanniny, yet fascinating feel of the cellars. Such projections are part of the myths about the world beneath the ground that have been inherited by Western culture and that came to lurk into Carlsberg's future urban plan.

To grasp landscape change as also the product of such cultural myths and ideas requires a close reading of human interpretation and how earlier interpretations and decisions can continue to influence development over a long period of time. Carlsberg's biography embodies several long-lived cultural phenomena. For example, the extensive celebration of Valby Hill occurs relatively persistently from the Bronze Age to the site's first industrial complex. Artists have romanticized the hill, too, as is illustrated in Vogt's choice of the hill as a central narrative for the future urban project. Fascination of topography is probably as old as human culture and is important in a relatively flat city like Copenhagen. Yet, despite of this apparent continuity, such inherited cultural traditions can give rise to very different views upon the specific landscape and their prospects. Although both the Bronze Age settlers and the baroque king may have agreed that Valby Hill is valuable as a symbolic and culturally elevated location, their ways of dealing with the hill differed as dramatically as from each other Vogt's use of it does from them.

Moreover, such inherited cultural ideas and myths are not always consistent. They can occur and disappear and be more or less strong in the encounter with other ideas and rationales in the shaping of the landscape in ways that can be described in the embeddedness or the dominant frames models. For example, the idea that the hill is a special, even unique, landscape form was not on the agenda for the planners of the plant after the 1930s, when the infrastructural rationale of trucks needing flat surfaces appeared and hence fought the hill. Then, in 2007-2008, the urban plan by Entasis quite literally pushed the slopes and terraces into small units in order to ensure large open squares filled with what was imagined as vivid city life. The year after, Vogt's plan made Valby Hill a main element in their landscape according to another idea of what urban space should be in the future and constructing a new and positive imaginary for the crisis-ridden urban development project. This shows how the accumulated interpretation as exemplified with Carlsberg's underground may continue over a long period of time, being reinterpreted, sometimes forgotten and reoccurring or growing in importance at certain points in time.

Language and changing practices of remembering and forgetting, changing aesthetic conceptions and ideas and processes of meaning-making are material to the way that urban landscapes are experienced, validated and redeveloped. To study change in this perspective gives an opportunity to reflect on the interpretation of sites and how it affects the way that places are represented and changed.

This chapter has shown three ways of understanding landscape change, more specifically the production of open spaces, deriving from Carlsberg's transformational history. Together they sketch a heuristic framework for a dynamic and relational way of understanding industrial open spaces beyond their standard perceptions. It should be noted that this framework is not exact science in a positivist way, though. While open space can be studied in a synchronic and diachronic perspective, the processes that produce space should be studied, as Lefebvre says, without being treated 'as a result or resultant, as an empirically verifiable effect of a past, a history or a society'.[15] Thus every biographical investigation, including the new chapters into Carlsberg's history offered in this book, is but a contribution to this infinite mesh of interpretation and reinterpretation.

15 Lefebvre, *The Production of Space*, 411.

Image accreditations

0.1 Photo: Adrian Täckmann

0.2 Aerial Photo: The Development Company the Carlsberg City

0.3 Entasis Architects visualization from 2009. The Development Company the Carlsberg City

0.4 Entasis Architects visualization of an urban space in the Carlsberg district. From the winning competition entry, 2007. The Development Company the Carlsberg City

0.5 The Development Company the Carlsberg City. Photo: Nana Reimers

0.6 Author's photo

0.7 Author's photo

0.8 The Development Company the Carlsberg City. Photo: unknown

0.9 Map from Lokalplan 432, City of Copenhagen. English translation by author

1.1 Author's photo

1.2 Author's photo

2.1 Author and Rikke Welan

2.2 Aerial Photo: The Development Company the Carlsberg City/COWI

2.3 Aerial Photo: The Development Company the Carlsberg City/COWI

2.4 Aerial Photo: The Development Company the Carlsberg City/COWI

2.5 Author's photo

2.6 Studio Egret West

2.7 Alberto Álvarez Agea, Ana Zazo Moratalla, Manuel Álvarez-Monteserín La Hoz, Ana Penalba Estebanez and Maria Mallo Zurdo

2.8. The Danish National Archives (Rigsarkivet), Forsvarets Arkiver, Militære kort og tegninger 1674-1980, mappe MBC-8, Danmark, Øvelser 1795, nr a.4.

2.9. Maps by author and Rikke Welan. Based on Bach, Bro, Mohr: *Gennem veje og alleer på Frederiksberg*

2.10	Painting by H.G.F. Holm, Museum of Copenhagen
2.11	Carlsberg Archives, Carlsberg Breweries Ltd.
2.12	Maps by author and Rikke Welan
2.13	Ny Carlsberg Glyptotek, photo: unknown
2.14	Carlsberg Archives, Carlsberg Breweries Ltd
2.15	Carlsberg Archives, Carlsberg Breweries Ltd
2.16	Carlsberg Archives, Carlsberg Breweries Ltd
2.17	Carlsberg Archives, Carlsberg Breweries Ltd
2.18	Drawing by Svenn Eske Kristensen, published in *Arkitektur* 14 1970, 'Udvidelse af tappehal for Carlsbergbryggerierne'
2.19	Author's photo
2.20	Author's photo
2.21	SLA
3.1	COWI/The Development Company the Carlsberg City
3.2	Author's photo
3.3	Author's photo
3.4	Author's photo
3.5	Author's photo
3.6	Author's photo
3.7	Author's photo
3.8	Carlsberg Archives, Carlsberg Breweries Ltd, drawing: unknown
3.9	Diagram by Svava Riesto with Rikke Welan
3.10	Entasis Architects/The Development Company the Carlsberg City/ the Municipality of Copenhagen
3.11	Entasis Architects
3.12	Photo collage: Jan Gehl/the Danish Architecture Press
3.13	The municipality of Copenhagen/The Development Company the Carlsberg City
3.14	Vogt Landscape Architects
3.15	Vogt Landscape Architects
3.16	Vogt Landscape Architects
3.17	Carlsberg Archives, Carlsberg Breweries Ltd, drawing: unknown
3.18	Carlsberg Ltd Properties/The Development Company the Carlsberg City, photo: unknown
3.19	Henning Larsen Architects and Domus
3.20	Henning Larsen Architects and Domus

3.21	Drawing by E. Erstad-Jørgensen in 1913-1914. Carlsberg Archives, Carlsberg Breweries A/S
3.22	Drawing by Rudolph Rothe, published in 1828 by Hartv. Fried. Dopps Bogtrykkerie, *Udtog af en Dagbog over Gartneri.*
3.23	Author's photo
3.24	Carlsberg Archives, Carlsberg Breweries Ltd, drawing: J.C. Jacobsen
3.25	Map author and Rikke Welan
3.26	Photo by Thommy Nilsen, Courtesy Carlsberg LTD Archive
3.27	Author's photo
3.28	Author's photo
3.29	Diagram author and Rikke Welan
3.30	Author's photos
3.31	Carlsberg Museum, Carlsberg Breweries Ltd, detail from painting by Franz Sedivy, 1926
3.32	Author's photo
3.33	Photo: Anders Hviid/Keingart
3.34	Photo: Anders Noerby/Keingart
3.35	Map: Author and Rikke Welan
3.36	Photo: unknown. Carlsberg Archives, Carlsberg Breweries A/S
3.37	Photo: unknown. Carlsberg Archives, Carlsberg Breweries A/S
3.38	Photo: unknown. Carlsberg Archives, Carlsberg Breweries A/S
3.39	Photo: unknown. Carlsberg Archives, Carlsberg Breweries A/S
3.40	Photo: unknown. Carlsberg Archives, Carlsberg Breweries A/S
3.41	Photo: Svend Rossen/UCC
3.42	Diagram by author and Rikke Welan, consulted by Ulla Nymand, Carlsberg Archives
3.43	Drawing by A. Fraenkel, published by H. Hagerups Boghandel in 1897, Carlsberg Archives, Carlsberg Breweries A/S
4.1	Entasis Architects/The Development Company the Carlsberg City
4.2	Entasis Architects/The Development Company the Carlsberg City
4.3	Entasis Architects/The Development Company the Carlsberg City
4.4	Author's photo
4.5	Entasis Architects/The Development Company the Carlsberg City

4.6	Diagram by author and Rikke Welan
4.7	Carlsberg Ltd Properties/The Development Company the Carlsberg City, photo: unknown
4.8	Carlsberg Ltd Properties/The Development Company the Carlsberg City, photo: unknown
4.9	Diagram by author and Rikke Welan, consulted by Ulla Nymand, Carlsberg Archives, and Nils Nielsen, Carlsberg Ltd Properties
4.10	Photo: abandon.dk
4.11	Photo: abandon.dk
5.1	Author's photo
5.2	Author's photo

References

Arbejdermuseet. 'Humlen ved Carlsberg.' http://humlenvedcarlsberg.dk/

Algreen-Ussing, Gregers, Grete Silding, and Allan de Wal. *Byens Træk. Om by-og bygningssystemet SAVE*. Copenhagen: Ministry of the Environment, 1992.

Appadurai, Arjun. 'Introduction: Commodities and the Politics of Value.' In *The Social Life of Things: Commodities in Cultural Perspective*, ed. Arjun Appadurai, 3-63. Cambridge: Cambridge University Press, 1986.

Ashworth, Gregory. 'Heritage and the Consumption of Places.' In *Bezeten van Vroeger. Erfgoed, Identiteit en Musealisering*, ed. R. van der Laarse, 193-206. Amsterdam: Het Spinhuis, 2005.

Bach, Michael, and Anne Møller-Rasmussen, eds. *Gennem Veje og Alléer på Frederiksberg*. Frederiksberg: Historisk-Topografisk Selskab for Frederiksberg, 2005.

Bakhtin, Mikhail. *The Dialogic Imagination: Four Essays*. Austin: University of Texas Press, 1994.

Bang, Herman. 'Et Besøg paa Carlsberg.' *Nationaltidende*, 29 May 1881.

Braae, Ellen. *Beauty Redeemed: Recycling Post-Industrial Landscapes*, Aarhus/Basel: Birkhäuser/Ikaros Press, 2015.

Braae, Ellen, and Lisa Diedrich. 'Site-Specificity in Contemporary Large-Scale Harbour Transformation Projects.' *Journal of Landscape Architecture* 7.1 (2012): 20-33.

Braae, Ellen and Riesto, Svava. 'As Found – a New Design Paradigm", *Nordic Journal of Architecture*, Nordic Journal of Architecture, Vol. 1, 2011, 8-9.

Braudel, Fernand. 'Histoire et Sciences sociales: La longue durée.' *Annales: Économies, Sociétés, Civilisations* 13.4 (1958): 725-753.

Brockwell, Sally, Sue O'Connor and Denis Byrne. *Transcending the Nature-Culture Divide in Cultural Heritage: Views from the Asia-Pacific Region*. Canberra: Anu E-Press, 2013.

Brorsen, Lisbeth (consultant at the Heritage Agency of Denmark). Personal communication. Copenhagen, 3 April 2008.

Bülow Jensen, Henriette. 'Humlebyen Arbejdernes Byggeforenings og kvarteret ved Vesterfælledvej, 1886-1911.' *Fabrik og Bolig* 2 (1997): 3-29.

Burns, Carol J., and Andrea Kahn, eds. *Site Matters: Design Concepts, Histories, and Strategies*. New York: Routledge, 2005.

Cameron, Christina, and Mechtild Rössler. *Many Voices, One Vision: The Early Years of the World Heritage Convention*. London: Ashgate, 2014.

Carlsberg Ltd. 'Carlsberg: Fredning er bureaukratisk og erhvervsfjendsk' (press release). Ritzau, 17 March 1997.

Certeau, Michel de. *The Practice of Everyday Life*. Berkeley: University of California Press, 1984.

Congrès International d'Architecture Moderne (CIAM). 'The Athens Charter' (1933). Trans. Anthony Eardley. In Le Corbusier, *The Athens Charter*. New York: Grossman, 1973.

Council of Europe. 'Council of Europe Framework Convention on the Value of Cultural Heritage for Society.' Council of Europe Treaty Series, No. 199. 27 October 2005. http://www.ecco-eu.org/fileadmin/user_upload/CETS_199.docxFARO_CONVENTION.pdf.

Cresswell, Tim. *Place: A Short Introduction*. London: Blackwell, 2004.

Cronon, William. *Nature's Metropolis: Chicago and the Great West*. New York: Norton, 1991.

Cullen, Gordon. *The Concise Townscape*. New York: Reinhold, 1961.

Danish Agency for Culture. 'How Does a Building Become Listed?' N.d. http://www.kulturstyrelsen.dk/english/cultural-heritage/listed-buildings/how-does-a-building-become-listed.

Dansk Bygningsarv and Realdania. *Industriens Bygningsarv.* http://bygningskultur2015.dk/industriarv/kortlaegning/.

Debord, Guy. 'Theory of the Dérive.' *Les Lèvres Nues* 9 (November 1956). Reprinted in *Internationale Situationniste* 2 (December 1958). English translation at http://www.cddc.vt.edu/sionline/si/theory.html.

Derfor kræver vi 6 timers arbejdsdag – nu! Rapport fra hverdagen, sommeren 1978 – arbejdere på Ny Tap Carlsberg. Copenhagen: Eget Forlag, 1979.

Development Company Carlsberg Byen. 'Carlsbergbyen Får Nye Ejere.' 11 April 2012. http://www.carlsbergbyen.dk/nyheder/carlsberg-byen-faar-nye-ejere/.

Development Company Carlsberg Byen. 'Realdaina Sælger sin Andel af Carlsbergbyen.' *Licitationen* 2. http://www.licitationen.dk/article/view/307506/realdania_saelger_sin_andel_af_carlsberg_byen#.

Dreyer, Kirsten. 'Kamma Rahbek og Johanne Langberg. Et bidrag til Bakkehusets historie.' In *Umisteligt. Festskrift til Erland Kolding Nielsen*, ed. John T. Lauridsen and Olaf Olsen, 227-251. Copenhagen: Det kongelige Bibliotek 2007.

Dümke, Cornelia, and Mikhail Gnedovsky. 'Social and Economic Value of Cultural Heritage.' European Expert Network on Culture, July 2013. http://www.interarts.net/descargas/interarts2557.pdf.

Edensor, Tim. *Industrial Ruins: Spaces, Aesthetics, Materiality.* New York: Berg, 2005.

Ellin, Nan. *Postmodern Urbanism.* Oxford: Blackwell, 1996.

Fabian, Louise, and Kristine Samson. 'Claiming Participation: A Comparative Analysis of DIY Urbanism in Denmark.' *Journal of Urbanism* 9.2 (2016): 166-184.

Fairclough, Graham. 'New Heritage Frontiers.' In *Heritage and Beyond*, ed. R. Palmer, 29-84. Strasbourg: Council of Europe, 2009.

Farias, Ignacio, and Thomas Bender. *Urban Assemblages: How Actor Network Theory Changes Urban Studies.* London: Routledge, 2010.

Fejerskov, Johan Moe. Carlsberg Byen lukker populære kulturtilbud på stribe, Politiken 24. October 2015.

Flyvbjerg, Bent. 'Five Misunderstandings about Case-Study Research.' *Qualitative Inquiry* 12.2 (2006): 219-245.

Forty, Adrian. *Words and Buildings: A Vocabulary of Modern Architecture.* London: Thames & Hudson, 2013.

Fowler, Peter. *The Past in Contemporary Society.* London: Routledge, 1992.

Fraenkel, A. *Gamle Carlsberg 1847-1897. Et bidrag til Dansk Industri Historie og Industriel Udviklingshistorie.* Copenhagen: H. Hagerups Boghandel, 1897.

Gaarmand, Arne. *Dansk byplanlægning 1938-1992.* Copenhagen: Danish Architectural Press, 1993.

Gandy, Matthew, *Concrete and Clay: Reworking Nature in New York City.* Cambridge, MA: MIT Press, 2002.

Gehl, Jan. *Livet mellem husene* [Life between buildings]. Copenhagen: Danish Architectural Press, 1971.

Gerritsen, Fokke. 'To Build and to Abandon: The Cultural Biography of Late Prehistoric Houses and Farmsteads in the Southern Netherlands.' *Archaeological Dialogues* 6 (1999): 78-114.

Glamann, Kristof. *Bryggeren. J.C. Jacobsen på Carlsberg.* Copenhagen: Gyldendal, 1996.

Glamann, Kristof. *The Carlsberg Foundation: The Early Years*. Copenhagen: Carlsberg Foundation, 2003.

Glamann, Kristof. *Øl og Marmor. Carl Jacobsen på Ny Carlsberg*. Copenhagen: Gyldendal, 1995.

Glamann, Kirsten, ed. *Din hengivne Jacobsen. Bryggerens breve til sin søn 1855-1870*. Copenhagen: Gyldendal, 1996.

Guttormsen, Torgrim Sneve, and Grete Swensen, eds. *Heritage, Democracy and the Public: Nordic Approaches*. London: Routledge, 2016.

Haddad, Elie. 'Christian Norberg-Schulz's Phenomenological Project in Architecture.' *Architectural Theory Review* 15.1 (2010): 88-101.

Harnow, Henrik, Keld Nielsen and Frank Allan Rasmussen, eds. 'The Museums and the Industrial Heritage: A Report from a Conference Held at Odense City Museums, 30 October-2 November 2003.' Copenhagen: Kulturarvsstyrelsen/Heritage Agency of Denmark, 2006. https://www.csu.edu/cerc/researchreports/documents/TheMuseumsAndTheIndustrialHeritage2006.pdf.

Harris, Steven, and Deborah Berke, eds. *Architecture of the Everyday*. New York: Princeton Architectural Press, 1997.

Harrison, Rodney. 'Beyond "Natural" and "Cultural" Heritage: Toward an Ontological Politics of Heritage in the Age of Anthropocene.' *Heritage & Society* 8.1 (2015): 24-42.

Hartmann, Sys. 'Fredning af industribygninger.' *Blød by* 43/44 (1988): 16-18.

Harvey, David. *Justice, Nature, and the Geography of Difference*. Cambridge, MA: Blackwell, 1996.

Hauser, Susanne, and Christa Kramleithner. *Ästhetik der Agglomeration*. Wuppertal: Müller und Busmann, 2006.

Hauxner, Malene. 'Carlsbergs landskab: et stykke kulturarv og kulturmiljø.' *Videnblade Planlægning af By og Land, Videntjenesten Skov og Landskab* 9.19-1 (2006): 1-2.

Hauxner, Malene. *Fantasiens Have. Det moderne gennembrud i havekunsten og sporene i byens landskab*. Copenhagen: Danish Architectural Press, 1993.

Hauxner, Malene. *Supernatur*. Aarhus: Ikaros Press, 2010.

Heimann Olsen, Erik. 'Carlsberg Bryggerierne imod fredning af sine gamle bryggerier.' In *Bygning, by og land*, 24-25. Holstebro: Landsforeningen for Bygnings- og Landskabskultur, 1997.

Hermansen, Tom, ed. 'Nye Byrum i København.' *Arkitektur DK* 5 (2012).

Heynen, Hilde, and Wessel de Jonge, eds. *Back from Utopia: The Challenge of the Modern Movement*. Rotterdam: 010 Publishers, 2002.

Hirschfeld, Christian Cay Lorenz. *Theorie der Gartenkunst*. Leibzig: Weidmann, 1779.

Hjort, Anders. 'Carlsbergbyens Pladser er blevet døbt.' *Politiken*, 24 November 2009.

Holst, Peter, and Sidsel Jacobsen. 'Visionen der forsvandt.' *Berlingske Tidene*, 17 October 2015.

Holtorf, Cornelius, and Graham Fairclough. 'The New Heritage and Re-shapings of the Past.' In *Reclaiming Archaeology: Beyond the Tropes of Modernity*, ed. Alfredo González Ruibal, 197-210. Routledge, 2013.

Hommels, Anique. *Unbuilding Cities: Obduracy in Urban Socio-Technical Change*. Cambridge, MA: MIT Press, 2005.

hooks, bell. *Yearning: Race, Gender, and Cultural Politics*. Boston: South End Press, 1990.

Hunt, John Dixon. *Greater Perfections: The Practice of Garden Theory*. London: Thames & Hudson, 2000.

Hvattum, Mari. 'Stedets Tyranni.' *Arkitektur* 4 (2010): 40-51.

Hyldtoft, Ole. 'Den teknologiske udvikling i Danmark.' In *Produktion og arbejdskraft i Danmark gennem 200 år*, ed. Flemming Mikkelsen, 37-58. Copenhagen: Nyt Fra Samfundsvidenskaberne, 1990.

Hyldtoft, Ole. *Københavns industrialisering 1840-1914*. Herning: Systime, 1984.

Ingold, Tim, *The Perception of the Environment. Essays on livelihood, dwelling and skill,* London and New York: Routledge, 2000

Ingold, Tim. *The Temporality of the Landscape,* World Archaeology Volume 25 No. 2, 152-174 1993

International Council on Monuments and Sites (ICOMOS). 'International Charter for the Conservation and Restoration of Monuments and Sites' (The Venice Charter). Second International Congress of Architects and Technicians of Historic Monuments, Venice, 1964. https://www.icomos.org/charters/venice_e.pdf.

Jackson, John Brinckerhoff. *Discovering the Vernacular Landscape.* New Haven: Yale University Press, 1984.

Jacobs, Jane. *The Death and Life of Great American Cities: The Failure of Town Planning.* New York: Vintage, 1961.

Jacobsen, J.C. 'Letter to Rudolph Rothe.' 24 June 1848. Rothes privatarkiv, Rigsarkivet/The Danish National Archives.

Jensen, Bernard Eric. *Kulturarv – et identitetspolitisk konfliktfelt.* Copenhagen: Gads Forlag, 2008.

Janssen, Joks; Luiten, Eric; Renes, Hans & Stegmeijer, Eva Heritage as sector, factor and vector: conceptualizing the shifting relationship between heritage management and spatial planning, European Planning Studies, (2017), 25:9, 1654-1672,

Jensen, Jørgen. *Danmarks Oldtid. Vol. 2, Broncealder 2.000-500 f.Kr.* Copenhagen: Nordisk Forlag, 2006.

Jørgensen, Caspar, ed. *Industri Industri. 25 Stk Dansk Kulturarv,* Copenhagen: Gads Forlag, 2007.

Jørgensen, Caspar. 'Carlsbergs bebyggelseshistorie. En industriarkæologisk skitse.' *Architectura* 31 (2009): 46-90.

Jørgensen, Caspar. *Carlsberg som nationalt industriminde.* Copenhagen: Heritage Agency of Denmark, 2006.

Jørgensen, Caspar. 'Changing Perceptions of Industrial Architecture in 1770-1940.' *Yhdyskun tasuunnittelu* 3 (2003): 16-32.

Jørgensen, Caspar. 'The Industrial Heritage in Denmark – Status 2003.' *Worklab Newsletter* 6 (2003): 75-83.

Jørgensen, Caspar. *Industribygninger i Københavns Kommune. Arbejdsnotat.* Copenhagen: Miljøministeriet/Planstyrelsen, 1991.

Jørgensen, Caspar. 'Industrisamfundets kulturarv – Kulturarvsstyrelsens første satsningsområde.' *Danske Museer* 3 (2003): 11-15.

Jørgensen, Caspar. 'Planning the Plant, 1845-1970: Debates and Realized Examples in Denmark, Sweden and Finland.' In *Industry and Modernism: Companies, Architecture, and Identity in the Nordic and Baltic Countries during the High-Industrial Period,* ed. Anja Kervanto Nevanlinna, 271-292. Helsinki: Studia Fennica Historica/The Finnish Literature Society, 2007.

Jørgensen, Caspar. 'Rational Planning as a Sign of Modernism.' In *Industry and Modernism: Companies, Architecture, and Identity in the Nordic and Baltic Countries during the High-Industrial Period,* ed. Anja Kervanto Nevanlinna, 66-95. Helsinki: Studia Fennica Historica/The Finnish Literature Society, 2007.

Jørgensen, Caspar. 'Tekno-økonomiske paradigmer og industrimiljøer I Danmark 1770-1970.' *Den Jyske Historiker* 121-122 (2009): 15-43.

Keiding, Martin, ed. *Arkitekturen på Carlsberg* [The architecture at Carlsberg]. Copenhagen: Danish Architectural Press, 2008.

Kirkwood, Niall, ed. *Remanufactured Sites: Rethinking the Post-Industrial Landscape.* London: Spon Press, 2001.

Kolen, Jan. 'De biografie van het landschap: Drie essays over landschap, geschiedenis en erfgoed.' PhD diss., VU University Amsterdam, 2005.

Kolen, Jan. 'Recreating (in) Nature, Visiting History: Second Thoughts on Landscape Reserves and Their Role in the Preservation and Experience of the Historic Environment.' *Archaeological Dialogues* 2 (1995): 127-159.

Kolen, Jan. 'Rejuvenation of the Heritage.' *'Scape* 2 (2006): 50-53.

Kolen, Jan, and Johannes Renes. 'Landscape Biographies: Key Issues.' In *Landscape Biographies: Geographical, Historical and Archaeological Perspectives on the Production and Transmission of Landscapes*, ed. Jan Kolen, Johannes Renes and Rita Hermans, 21-47. Amsterdam: Amsterdam University Press, 2015.

Kolen, Jan, and Mathijs Witte. 'A Biographical Approach to Regions, and Its Value for Spatial Planning.' In *Multiple Landscape: Merging Past and Present*, ed. Wim van der Knaap and Arnold van der Valk, ?-?. Wageningen: International Study Group on Multiple Use of Land ISOMUL, 2006.

Kolen, Jan, Johannes Renes and Rita Hermans, eds. *Landscape Biographies: Geographical, Historical and Archaeological Perspectives on the Production and Transmission of Landscapes*. Amsterdam: Amsterdam University Press, 2015.

Kommuneatlas Viborg, Bevaringsværdier i byer og bygninger, Planstyrelsen, Viborg commune, Elgstrømgruppen Aps. Copenhagen: Hans Reitzels Forlag, 1990.

Koolhaas, Rem. 'Generic City.' In *Small, Medium, Large, Extra-Large: Office for Metropolitan Architecture*, by Rem Koolhaas and Bruce Mau, and ed. Jennifer Sigler, 1248-1257. Rotterdam: 010 Publishers, 1995.

Kopytoff, Igor. 'The Cultural Biography of Things: Commoditization as Process.' In *The Social Life of Things: Commodities in Cultural Perspective*, ed. Arjun Appadurai, 64-91. Cambridge: Cambridge University Press, 1986.

Kristensen, Svenn Eske. *Strejftog. Et liv i byggeriets tjeneste*. Copenhagen: Eget Forlag, 1977.

Kristensen, Svenn Eske. 'Udvidelse af tappehal for Carlsberg Bryggerierne.' *Arkitektur* 14 (1970): 224-227.

Kruse, Tove. *Bryggeriarbejdere i København – de første årtier*. Roskilde: Roskilde University, 1992.

Kulturarv. En værdifuld ressource for kommunernes udvikling. En analyse af danskernes forhold til kulturarv. Copenhagen: Heritage Agency of Denmark and Realdania, 2005.

Kulturarven – et aktiv. Erfaringer fra de danske kulturarvskommuner. Copenhagen: Heritage Agency of Denmark and Realdania, 2007.

Larsen, Carsten Ulrik. *Gravhøje på Sjælland, Nationalmuseets Arbejdsmark*. Copenhagen: National Museum, 1989,

Lefebvre, Henri. *The Production of Space*. Trans. Donald Nicholson-Smith. Oxford: Blackwell, 1991.

Lefebvre, Henri. *The Urban Revolution*. Trans. Robert Bononno. Minneapolis: University of Minnesota Press, 2003.

Lichtenstein, Claude, and Thomas Schregenberger, eds. *As Found: The Discovery of the Ordinary*. Zürich: Lars Müller Publishers, 2001.

Loewental, David. *The Heritage Crusade and the Spoils of History*. Cambridge: Cambridge University Press, 1997.

Loudon, John Claudius. *An Encyclopaedia of Cottage, Farm, and Villa Architecture and Furniture: Containing Numerous Designs for Dwellings, from the Cottage to the Villa*. London: Longman, 1833.

Lund Jørgensen, Mette. *Valby Bakke*. Copenhagen: Rhodos, 2009.

Lynch, Kevin. *What Time Is This Place?* Cambridge, MA: MIT Press, 1972.

Magnani, Lauro. 'Genoese Gardens: Between Pleasure and Politics.' In *Gardens, City Life and Culture: A World Tour*, ed. Michel Conan and Wangheng Chen, 55-73 Washington, DC: Dumbarton Oaks Research Library and Collection, 2008.

Malpas, Jeff. *Place and Experience: A Philosophical Topography*. Cambridge: Cambridge University Press, 1999.

Markmann, Anders. 'Tegl i industribyggeri. Udvidelse af Carlsberg Bryggeriernes aftapningshal.' *Tegl* 3 (1960): 225-228.

Markussen, Ib, and Niels Kelstrup. 'Bryggerens Vand.' *Geologi.Nyt fra GEUS* (1997): 1-4.

Marot, Sébastien. 'The Reclaiming of Sites.' In *Recovering Landscape: Essays in Contemporary Landscape Architecture*, ed. James Corner, 45-66. New York: Princeton Architectural Press, 1999.

Marot, Sébastien. Sub-urbanism and the Art of Memory. London: AA Publications, 2003.

Marot, Sébastien. 'Sub-urbanism/Super-urbanism. From Central Park to La Villette.' *AA Files* 53 (2006): 21-37.

Massey, Doreen. 'A Global Sense of Place.' In *Reading Human Geography*, ed. T. Barnes and D. Gregory, 315-323. London: Arnold, 1997.

Mattingly, C. 'Narrative Reflections on Practical Actions: Two Learning Experiments in Reflective Storytelling.' In *The Reflective Turn: Case Studies In and On Educational Practice*, ed. D.A. Schön, 235-257. New York: Teachers College Press, 1991.

McHarg, Ian. *Designing with Nature*. Garden City: Natural History Press, 1969.

Meinig, Donald William, ed. *The Interpretation of Ordinary Landscapes: Geographical Essays*. New York: Oxford University Press, 1979.

Meyer, Karl. *De Forenede Bryggerier 1891-1916*. Copenhagen: De Forenede Bryggerier, 1916.

Michaelsen, M.K. 'Carlsbergbryggeriernes Bygningshistorie.' *Arkitektens Månedshæfte*, August-September (1929): 135-192.

Miljøvurdering og VVM for Carlsberg Valby – 'Vores By'. Teknisk Rapport. Copenhagen: Københavns Kommune, Carlsberg Ltd Properties and COWI Ltd, 2008.

Ministry of the Environment. *SAVE: Kortlægning og registrering af byers og bygningers bevaringsværdi og udarbejdelse af kommuneatlas. Vejledning*. Copenhagen: Ministry of the Environment, 1997.

Muratori, Saviero. *Studi per una operante storia urbana di Venezia*. Rome: Instituto Poligrafico dello Stato, 1959.

Netherlands State Government. *Nota Belvedere: Beleidsnota over de relatie cultuurhistorie en ruimtelijke inrichting* [The Belvedere Memorandum: A policy document examining the relationship between cultural history and spatial planning]. Gravenhage, 1999.

Nielsen, Niels (head of projects, Carlsberg Ltd Properties/The Development Company Carlsbergbyen). Personal communication. Copenhagen, 2 October 2009.

Nielsen, Rolf. *Gamle Carlsberg 1847-1897. Fra Håndværk til Stordrift*. Copenhagen: Carlsberg Bryggerierne, 1984.

Nielsen, Tom. *Gode Intentioner og Uregerlige Byer*. Aarhus: Arkitektskolens Forlag, 2010.

Norberg-Schulz, Christian. *Genius Loci. Towards a Phenomenology of Architecture*. London: Academy Editions, 1980.

Nørballe, Inge, Anne Clante and Jørgen Fisker. *Omkring Bakkehuset, Bakkehuset og dets beboere set med samtidens øjne*. Frederiksberg: Fisker & Schou, 1993.

Nymand, Ulla (head, Carlsberg Ltd Archive). Interview. 7 July 2011.

Nymand, Ulla (head, Carlsberg Ltd Archive). Personal communication. Copenhagen, 31 October 2008.

Nymand, Ulla (head, Carlsberg Ltd Archive). Personal communication. Copenhagen, 7 July 2010.
Nystrøm, Eiler. *Frederiksbergs Historie I-III.* Copenhagen: Gyldendal, 1942-1956.

Palmer, Marilyn, and Peter Neaverson. *Industrial Archaeology: Principles and Practice.* London: Routledge, 1998.
Paludan-Müller, Carsten. 'Kulturarv: Forskning og forvaltning. Generelle tendenser i Europa.' *Arkæologisk Forum* 9 (2003): 7-12
Pedersen, Steen Estvald. *Bryggerens Akademi.* Copenhagen: Carlsberg Foundation, 2001.
Prominski, Martin. *Landschaft entwerfen: Zur Theorie aktueller Landschaftsarchitektur.* Berlin: Reimer, 2004.

Raskin, Laura. 'Jorge Otero-Pailos and the Ethics of Preservation.' *Places Journal,* January 2011. https://placesjournal.org/article/jorge-otero-pailos-and-the-ethics-of-preservation/.
Rasmussen, Steen Eiler. *København. Et bysamfunds særpræg og udvikling gennem tiderne.* Copenhagen: Gads Forlag, 2001.
Relph, Edward. *Place and Placelessness.* London: Pion, 1976.
Renes, Johannes. 'Layered Landscapes.' In *Landscape Biographies: Geographical, Historical and Archaeological Perspectives on the Production and Transmission of Landscapes,* ed. Jan Kolen, Johannes Renes and Rita Hermans, 403-421. Amsterdam: Amsterdam University Press, 2015.
Riegl, Alois: *Der moderne Denkmalkultus, sein Wesen, seine Entstehung* , Vienna:. Tr. K. W. Forster and D. 1903.Ghirardo
Riesto, Svava. 'The Old, the Pure, and the Quirky: Contested Heritage Values in the Urban Redevelopment of the Carlsberg Breweries in Copenhagen.' In *A Critical Biographical Approach of Europe's Past,* ed. Dirk Callebaut, 111-125. Gent: Provinciebestuur Oost-Vlaanderen, 2016.
Riesto, Svava, and Anne Tietjen. 'Doing Heritage Together: New Heritage Frontiers in Participatory Planning.' In *Heritage, Democracy and the Public: Nordic Approaches,* ed. Torgrim Sneve Guttormsen and Grete Swensen, 159-175. London: Routledge, 2016.
Riesto S, Søberg M, Braae EM. 'City of open works', in I Diedrich L, Hendriks M, Kandjee T, Moll C, ed., In Touch. Basel: Birkhäuser Verlag GmbH. 2012. 237-240.
Riesto, Svava. 'Finding Industrial Space: Contested Spatial Concepts for Carlsberg Square, Copenhagen.' *Nordic Journal of Architecture* 1.1 (2011): 30-39.
Riesto, Svava, and Malene Hauxner. 'Underliggende Grunde. Om Carlsbergs kældre og byomdannelse som kulturel aktivitet.' *Arkitekten* 113 (2011): 95-99.
Riesto, Svava. '(Hvordan) skal vi huske Christian Norberg-Schulz?' *Byggekunst* 3/4 (2001): 18-23.
Rifbjerg, Synne. 'Brygge og Bygge.' *Weekendavisen,* 23 January 2009.
Rosenvinge, Line. 'Jeg drikker aldrig mere en Carlsberg.' *Børsen,* 2 December 2015.
Roskilde – bevaringsværdier i byer og bygninger. Planstyrelsen, Roskilde Kommune, Elgstrømgruppen ApS. Copenhagen: Hans Reitzels Forlag, 1990.
Rothe, Rudolph. 'Bemærkning til Planen over Ordningen af Omgvivelserne ved Bryggeriet ved Valby 1848 (Jacobsen).' Copenhagen, 1 February 1848. Rothes privatarkiv, Rigsarkivet/The Danish National Archives.
Rothe, Rudolph. *Landskabsgartneriske Betragtninger over Danmark.* Copenhagen: Gyldendalske Boghandling, 1853.
Rothe, Rudolph. *Udtog af en Dagbog Over Gartnerie, især med Hensyn til Driverier og den skjønne Haugekunst,ført paa en Rejse igjennem Tyskland, Over-Italien, Frankrige og Holland af Rudolph Rothe.* Copenhagen: Hartv. Fried. Dopps Bogtrykkerie, 1828.
Rowe, Colin, and Fred Koetter. *Collage City.* Cambridge, MA: MIT Press, 1978.

Roymans, Nico. 'The Cultural Biography of Urnfields and the Longterm History of a Mythical Landscape.' *Archaeological Dialogues* 2 (1995): 2-38.

Roymans, Nico, Fokke Gerritsen, Cor Van der Heijden, Koos Bosma and Jan Kolen. 'Landscape Biography as Research Strategy: The Case of the South Netherlands Project.' *Landscape Research* 34.3 (2010): 337-359.

Roymans, Nico, Fokke Gerritsen, Cor Van der Heijden, Koos Bosma and Jan Kolen. 'Revitalizing History: Moving from Historical Landscape Reconstructions to Heritage Practices in the Southern Netherlands.' In *The Cultural Landscape and Heritage Paradox*, ed. Tom Bolemers, Henk Kars, Arnold Van der Valk, Mies Viejnen, 387-406 Amsterdam: Amsterdam University Press, 2010.

Samuels, Marwyn S. 'The Biography of Landscape: Cause and Culpability.' In *The Interpretation of Ordinary Landscapes: Geographical Essays*, ed. Donald William Meinig, 51-88. New York: Oxford University Press, 1979.

SAVE: Kortlægning og registrering af byers og bygningers bevaringsværdi og udarbejdelse af kommuneatlas. Vejledning. Copenhagen: Ministry of the Environment, 1997.

Schmid, Christian. 'Henri Lefebvre's Theory of the Production of Space: Towards a Three-Dimensional Dialectic.' In *Space, Difference, Everyday Life: Reading Henri Lefebvre*, ed. Kanishka Goonewardena, Stefan Kipfer, Richard Milgrom and Christian Schmid, 27-46. Routledge: London, 2008.

Schmid, Christian. *Stadt, Raum und Gesellschaft. Henri Lefebvre und die Theorie der Produktion des Raumes.* Stuttgart: Franz Steiner Verlag, 2010.

Schön, Donald. *The Reflective Practitioner: How Professionals Think in Action.* London: Temple Smith, 1983.

Seamon, David, and Jacob Sowers. 'Place and Placelessness: Edward Relph.' In *Key Texts in Human Geography*, ed. P. Hubbard, R. Kitchen and G. Vallentine, 43-51. London: Sage, 2008.

Seggern, Hille von, Julia Werner and Lucia Grosse-Bächle. *Creating Knowledge: Innovation Strategies for Designing Urban Landscapes.* Berlin: Jovis, 2008.

Selmer, Jørgen, ed. *Industrisamfundets Kulturarv. Regional indberetning af industriminder i Københavnsregionen i henhold til Kulturarvsstyrelsens brev j.nr. 2003-300-0001 i forbindelse med satsningsområdet 'Industrisamfundets Kulturarv'.* Copenhagen: National Museum et al., 2007.

Sestoft, Jørgen. *Danmarks Arkitketur. Arbejdets bygninger.* Copenhagen: Gyldendalske Boghandel/Nordisk Forlag, 1985.

Sestoft, Jørgen, and Kell Elgstrøm. *Industribygninger og deres bevaring.* Ministry of the Environment, Fredningsstyrelsen, 1985.

Sieverts, Thomas. *Cities without Cities: An Interpretation of the Zwischenstadt.* London: Spon Press, 2003.

Sieverts, Thomas. 'The Prerequisite of a Sustainable City Is Beauty.' Interview with Thomas Sieverts by Søren Schmidt-Jensen, Danish Centre for Architecture/Sustainable Cities, 21 January 2014. http://www.dac.dk/en/dac-cities/sustainable-cities/experts/thomas-sieverts-the-prerequisite-of-a-sustainable-city-is-beauty/.

Sieverts, Thomas. 'Von der unmöglichen Ordnung zu einer möglichen Unordnung im Entwerfen der Stadtlansdchaft.' *DISP* 169.2 (2007): 5-16.

Sieverts, Thomas. 'Where We Live Now.' Trans. Diana George. In *Where We Live Now: An Annotated Reader*, ed. Matthew Stadler, 21-81. New York: Suddenly, 2008.

Sieverts, Thomas. *Zwischenstadt: Zwischen Ort und Welt, Raum und Zeit, Stadt und Land.* Basel: Birkhäuser, 1999.

Smith, Laurajane. *Uses of Heritage*. London: Routledge, 2006.

Solà-Morales, Manuel de. 'The Culture of Description.' *Perspecta* 25 (1989): 16-25.

Stake, Richard. *The Art of Case Study Research*. Thousand Oaks: Sage, 1995.

Stanek, Lukasz. *Henri Lefebvre on Space: Architecture, Urban Research, and the Production of Theory*. Minneapolis: University of Minnesota Press, 2011.

Steenstrup, Johannes. *Carl Jacobsens Liv og Gerning*. Carlsberg: Ny Carlsbergfondet, 1922.

Steenstrup, Johannes. *Carl Jacobsen og hans Fader. En skildring af to danske borgeres udvikling*. Copenhagen: M.P. Madsen, 1921.

Stenak, Morten, ed. *SAVE. Kortlægning og registrering af bymiljøers og bygningers bevaringsværdi*. Copenhagen: Heritage Agency of Denmark, 2011.

Storm, Anna. *Hope and Rust: Reinterpreting the Industrial Place in the Late 20th Century*. Stockholm: Kungliga Tekniska Högskolan, 2008.

Strømstad, Poul, ed. *J.C. Jacobsens Bryggergård i Brolæggerstræde i 75-året for dets stiftelse*. Copenhagen: Forum, 1977.

Tietjen, Anne. *Towards an Urbanism of Entanglement*. Aarhus: Arkitektskolens Forlag, 2011.

Tietjen, Anne, Svava Riesto and Pernille Skov, eds. *Forankring i forandring Christiania og bevaring som ressource i byomdannelse*. Aarhus: Arkitektens Forlag, 2007.

Toft Jensen, Hannelene. *Vesterbro Bydelsatlas. Bevaringsværdier i Bydel og Bygninger*. Copenhagen: Ministry of the Environment, 1991.

Trindler, Barry. *The Making of the Industrial Landscape*. London: Phoenix Giant, 1997.

Tschudi-Madsen, Stefan. *Restoration and Anti-Restoration: A Study in English Restoration Philosophy*. Oslo: Universitetsforlaget, 1976.

Tullberg, Poul. *Dansk Bryggerifortegnelse*. Silkeborg: Skandinavisk Bryggerisouvenir Samlerforening, 1995.

Tunbridge, John E., and Gregory Ashworth. *Dissonant Heritage: The Management of the Past as a Resource in Conflict*. Chichester: Wiley, 1996.

Tyrwhitt, Jacqueline, José Luis Sert and Ernesto Nathan Rogers, eds. *The Heart of the City: Towards the Humanisation of Urban Life*. London: Lund Humphries, 1952.

Veldhoven, Felix. 'Post-Industrial Coal-Mining Landscapes and the Evolution of Mining Memory.' In *Landscape Biographies: Geographical, Historical and Archaeological Perspectives on the Production and Transmission of Landscapes*, ed. Jan Kolen, Johannes Renes and Rita Hermans, 327-339. Amsterdam: Amsterdam University Press, 2015.

Venturi, Robert, Denise Scott Brown and David Izenour. *Learning from Las Vegas*. Cambridge, MA: MIT Press, 1994.

Venturi, Robert, Denise Scott Brown and David Izenour. *Signs of Life: Symbols in the American City*. New York: Aperture, 1976.

Vindum, Kjeld. 'Harild på Carlsberg.' *Scala* 28 (1993): 42-45.

Vogt, Günther. *Miniature and Panorama*. Baden: Lars Müller Publishers, 2006.

Voss, Ralph Günter (landscape architect, Vogt Landscape Architects). Personal communication. Copenhagen, 15 February 2009.

While, Aidan, and Michael Short. 'Place Narratives and Heritage Management: The Modernist Legacy in Manchester.' *Area* 43.1 (2011): 4-13.

Williams, Rosalind. *Notes on the Underground. An Essay on Technology, Society, and the Imagination*. Cambridge, MA: MIT Press, 2008.

Willim, Robert. 'Industrial Cool – om det materiellas närvaro och industri samhällets frånvaro.' In *Materialitet og Dannelse*, ed. Minna Kragelund and Lene Otto, 159-174. Copenhagen: Danmarks Pædagogiske Universitet, 2005.
Wilson, Chris, and Paul Groth, eds. *Everyday America: Cultural Landscape Studies after J.B. Jackson*. London: University of California Press, 2003.

Zanker von Meyer, Dorothea. *Die Bauten von J.C. und Carl Jacobsen. Zur Bautätigkeit einer Industriellenfamilie in Dänemark*. Munich: Deutscher Kunstverlag, 1982.
Zukin, Sharon. *Loft Living: Culture and Capital in Urban Change*. New York: Radius, 1988.
Zukin, Sharon. *Naked City. The Death and Life of Authentic Urban Places*. New York: Oxford University Press, 2011.

Design proposals, plans and documents from the Carlsberg redevelopment

Alberto Álvarez Agea, Ana Zazo Moratalla, Manuel Álvarez-Monteserín Lahoz, Ana Peñalba Estébanez and María Mallo Zurdo. 'Time Will Tell.' Competition entry for the international ideas competition 'Carlsberg Vores By' (second prize), 2007.
Arkitektforening [Danish Association of Architects], Jens Kvorning and Helle Søholt. 'Carlsberg Vores By: Open International Ideas Competition: Jury Report.' Copenhagen: Carlsberg Ltd Properties, 2007.
Carlsberg Ltd Properties. 'Vores By Carlsberg: Open International Ideas Competition: Brief.' In consultation with Arkitektforening (Danish Association of Architects), the City of Copenhagen, and the Heritage Agency of Denmark. Copenhagen, 2006.
Carlsberg Ltd Properties and The Heritage Agency of Denmark, 'Arven på Carlsberg. Eksisterende Bygninger på Carlsberg. *Arbejdsdokument,* Copenhagen:Carlsberg Ltd Properties and The Heritage Agency of Denmark, april 2008.
Copenhagen Convention Bureau. 'Awards & Accolades for Copenhagen' (section: 'Lifestyle, liveability and design'). http://www.copenhagencvb.com/copenhagen/awards-accolades-copenhagen.
Copenhagen Municipality. *Arkitekturby København. Københavns Kommunes Arkitekturpolitik*. Copenhagen: Copenhagen Municipality, 2010.
Copenhagen Municipality. 'Copenhagen Urban Space Action Plan' (special edition in English for the Sixth Biennial of Towns and Town Planning, June 2006). Copenhagen, 2005.
Copenhagen Municipality. 'Lokalplan 44, Carlsberg Bryggerier.' Copenhagen: Copenhagen Municipality, 1983.
Copenhagen Municipality. 'Lokalplan 432, Carlsberg II.' Centre for Bydesign, Copenhagen, 18 February 2009.
Entasis Architects. 'Vores Rum.' Competition entry for the international ideas competition 'Carlsberg Vores By' (first prize). Carlsberg Ltd Properties, February 2008.
Entasis Architects. 'Vores Rum: Plan Elaboration upon Competition Entry.' 2 vols. Carlsberg Ltd Properties, February 2008.
Henning Larsen Architects and Domus Architects. 'Carlsberg Bryggerierne. Fysisk Udviklingsplan. Forudsætninger og Berergninger.' Carlsberg Ltd, 1993.
Heritage Agency of Denmark. 'Carlsberg som nationalt industriminde.' Communication to Carlsberg Ltd Properties, 22 September 2006.

Ministry of Culture and the Heritage Agency of Denmark. 'Udvidelse af fredningerne på Carlsberg.' Communication to Carlsberg Ltd, 4 March 2009.

Ministry of the Environment and Energy. 'Fredningsbeslutning Matr. Nr. 20a, Valby Kvarter, København.' Communication to Carlsberg Ltd, 15 November 1999.

Studio Egret West. 'Collage Village.' Competition entry for the international ideas competition 'Carlsberg Vores By,' 2007.

Ulvemann Explorative and Carlsberg Ltd Properties. 'En kvalitativ undersøgelse vedr. image af Carlsberg og Apollo.' Copenhagen, 2006.

Vogt Landscape Architects. 'Bryghuspladsen/The Carlsberg Square.' Carlsberg Ltd Propertie, June 2009.

Vogt Landscape Architects. 'The Story of the Hill.' Carlsberg Ltd Properties, May 2010.

For Product Safety Concerns and Information please contact our EU representative GPSR@taylorandfrancis.com Taylor & Francis Verlag GmbH, Kaufingerstraße 24, 80331 München, Germany

Batch number: 10722100

Printed by Printforce, the Netherlands